SCIENCE
Life Science
Earth Science
Chemistry
Physics

Susan D. McClanahan, Educational Consultant
Donna D. Amstutz, Special Advisor

STECK-VAUGHN ADULT EDUCATION ADVISORY COUNCIL

Donna D. Amstutz
Asst. Project Director
Northern Area Adult Education Service Center
Northern Illinois University
DeKalb, Illinois

Roberta Pittman
Director, Project C3 Adult Basic Education
Detroit Public Schools
Detroit, Michigan

Elaine Shelton
Consultant, Competency-Based Adult Education
Austin, Texas

Lonnie D. Farrell
Supervisor, Adult Special Programs
Los Angeles Unified School District
Los Angeles, California

Don F. Seaman
Professor, Adult Education
College of Education
Texas A&M University
College Station, Texas

Bobbie L. Walden
Coordinator, Community Education
Alabama Department of Education
Montgomery, Alabama

Meredyth A. Leahy
Director of Continuing Education
Cabrini College
Radnor, Pennsylvania

Jane B. Sellen
Supervisor, Adult Education
Western Iowa Tech Community College
Sioux City, Iowa

A Subsidiary of National Education Corporation

Product Design and Development: McClanahan & Company with PC&F, Inc.
Project Director: Bonnie Diamond, Ed. D.
Assistant Project Director: Patricia Carlin
Design/Production Director: Judi Baller
Editorial Development: David Steinecker, Marcia Mungenast, Bunny Gorman

Photograph Credits: Cover Photograph by Rick Patrick
Photoresearch by Photosearch, Inc.
- pg. 12   S. J. Krasemann, Peter Arnold
- pg. 78   George Gerster, Photo Researchers
- pg. 87   NASA
- pg. 94   bottom: Michal Heron, Woodfin Camp
- pg. 101 United States Geological Survey
- pg. 110 Albert Steiner, Photo Researchers
- pg. 128 Heinz Kluetmeier, Sports Illustrated
- pg. 137 General Electric Research Laboratory

Illustration Credits: PC&F, Inc.

Copyright © 1988 by Steck-Vaughn Company. All rights reserved. No part of this publication may be reproduced or transmitted in any form or by any means, electronic or mechanical, including photocopy, recording, or any information storage and retrieval system, without permission in writing from the publisher. Requests for permission should be addressed to Permissions, Steck-Vaughn Company, P.O. Box 2028, Austin, TX 78768.
Printed in U.S.A.

ISBN 0-8114-1895-2

Contents

To the Student,
How To Use This Book,
Test-taking and Study Skills   1–3

PRETEST
Answers and Explanations   4–11

LIFE SCIENCE OVERVIEW   12–13

  1. **The Biology of Cells**   14–20
    STRATEGY: Identify the Main Idea   15

  2. **Photosynthesis**   21–27
    STRATEGY: Restate Information   22

  3. **Evolution**   28–34
    STRATEGY: Identify Cause and Effect Relationships 29

  4. **Plant Growth**   35–41
    STRATEGY: Identify an Implication   36

  5. **The Human Digestive System**   42–48
    STRATEGY: Recognize Unstated Assumptions   43

  6. **Defenses Against Disease**   49–55
    STRATEGY: Recognize the Role of Values and Beliefs 50

  7. **Mendelian Genetics**   56–62
    STRATEGY: Assess Appropriateness of Data to Prove or Disprove
      Statements   57

  8. **Ecosystems**   63–69
    STRATEGY: Identify Logical Fallacies in Arguments 64

LIFE SCIENCE REVIEW   70–77

EARTH SCIENCE OVERVIEW   78–79

  9. **The Planet Earth**   80–86
    STRATEGY: Identify an Implication   81

  10. **Air and Water**   87–93
    STRATEGY: Use Given Ideas in Another Context   88

  11. **The Earth's Resources**   94–100
    STRATEGY: Distinguish Fact From Opinion   95

  12. **The Changing Earth**   101–107
    STRATEGY: Identify Cause and Effect Relationships 102

EARTH SCIENCE REVIEW	108–109

CHEMISTRY OVERVIEW — 110–111

13. Matter — 112–118
  STRATEGY: Use Given Ideas in Another Context — 113

14. Chemical Reactions — 119–125
  STRATEGY: Assess Adequacy of Data to Support Conclusions — 120

CHEMISTRY REVIEW — 126–127

PHYSICS OVERVIEW — 128–129

15. Motion, Thermodynamics — 130–136
  STRATEGY: Distinguish Conclusions from Supporting Statements — 131

16. Electricity and Magnetism — 137–143
  STRATEGY: Identify Cause and Effect Relationships — 138

17. Waves — 144–150
  STRATEGY: Assess Adequacy of Data to Support Conclusions — 145

PHYSICS REVIEW — 151–152

ANSWERS and EXPLANATIONS for REVIEWS — 153–158

Answer Sheet — 159–160

POSTTEST

Answers and Explanations — 161–167

PRETEST/POSTTEST DIAGNOSTIC CHART — 168

Index — 169–172

Diagrams 5, 6, 14, 16, 21, 24, 25, 26, 28, 31, 33, 35, 42, 44, 46, 49, 51, 58, 60, 70, 71, 72, 73, 74, 75, 80, 87, 90, 99, 106, 108, 119, 124, 127, 133, 140, 148, 161, 162, 163, 164

Tables 58, 83, 117

Charts 4, 37, 46, 58, 66, 76

To the Student

The GED test offers you an opportunity to

1. Keep or get a better job in government, industry or business
2. Increase your earning powers
3. Expand educational opportunities in trade, technical, vocational, or apprenticeship programs
4. Fulfill your personal goals

It awards a certificate that is the equivalent of a high school diploma. It measures your mastery of skills and general knowledge in Writing Skills, Social Studies, Science, Reading Literature and the Arts and Mathematics.

The Steck-Vaughn program prepares you for success on the GED exam by:

- Teaching appropriate concepts and skills that will provide a solid foundation for your general knowledge
- Providing practice in the GED format
- Emphasizing the reading skills that will be tested on the GED exam—those that require you to apply, to analyze and to evaluate as well as to comprehend what you read
- Offering test-taking tips to build your confidence
- Applying concepts and skills in practical and realistic settings
- Building vocabulary by highlighting and defining new terms
- Teaching reading, writing and problem-solving skills to make you better readers
- Frequently using charts, tables, graphs, diagrams, maps and figures, which are part of the GED test, and instructing you how to gain meaning from them.

It does this through an easy-to-follow, predictable Four Step Plan that includes

- Introducing and teaching a concept
- Applying a particular reading or problem-solving strategy to the concept
- Practicing the concept in the GED test format
- Testing and checking your answers

(Answers and Explanations to both the Practice and GED Mini-Test items provide further instruction through explanations of why choices are incorrect as well as why a given choice is correct.)

The following table summarizes the contents of the GED tests.

The Tests of General Educational Development

Test	Content Areas	Number of Items	Time Limit (minutes)
Writing Skills	PART ONE Sentence Structure Usage Mechanics	55	75
Writing Sample	PART TWO Essay	1	45
Social Studies	United States History Geography Economics Political Science Behavioral Science	64	85
Science	Life Science Earth Science Physics Chemistry	66	95
Reading Literature and the Arts	Popular Literature Classical Literature Commentary	45	65

Mathematics	Arithmetic
	  Measurement
	  Number
	    Relationships
	  Data Analysis
	Algebra
	Geometry
		56	90

How To Use This Book

The Book:
A sequentially organized program

The Pretest:
Tells you what content and skills you have already mastered and What content and skills you need to work on—a real aid to planning your time and increasing your studying efficiency

The Overview:
- Explains each major section of the book
- Provides definitions of vocabulary terms and concepts that relate to material in each major section

The Study Plan:

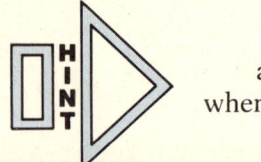 a practical reminder when applying a concept

 a practical reminder related to test-taking

A predictable seven-page lesson that includes:
- An **Introductory,** teaching page
- A **Strategy** page that teaches and applies a related reading skill to increase understanding and aid mastery
- **Practice** pages that review and reinforce the particular content and the reading skill in GED format
- **Mixed Practice** exercises in the Mathematics and Writing texts that review previously learned material
- A multiple choice, GED format test (**GED Mini-Test**) that measures higher-level thinking skills
- **Answers and Explanations** for both the Practice and the GED Mini-Tests that give *immediate* feedback and pinpoint possible errors or weaknesses

The Review:
- Summarizes the instructional content of a section
- Provides more practice items in GED format

The Posttest:
- Simulates the actual GED test
- Alerts you to the need for possible further study

Test-Taking and Study Skills

Test-Taking Skills

The AIM of the Steck-Vaughn GED program is to prepare you to take and to pass the GED examination with ease and confidence. You bring to the program your own personal style and your life experience.

With these as a base, use the preparation material and the suggestions that follow to build and strengthen your academic skills, test-taking ability and study skills.

The Steck-Vaughn program is designed to provide numerous test-taking situations in the multiple-choice GED format. GED items appear in:

- Pre- and posttests for each book
- Practice and test pages for each lesson
- Reviews for each major book section

The more opportunities you have to practice the GED test format, the more you will increase your confidence in test taking. *You learn about the test by preparing for the test.*

Some key test-taking skills are:

- Set goals
- Plan your test-taking time
- Read for understanding
- Analyze the test questions carefully before answering them
- Pace yourself

Reviewing the helpful Steck-Vaughn GED Mini-Test Tips that are part of each lesson will help you gain more confidence in test taking.

Study Skills

The Steck-Vaughn program aids in developing and improving your study skills. They are an important element in successful test taking.

Some important study skills to remember are

- Improve your vocabulary
- Use your text as well as other resources, including maps, charts, graphs, diagrams to help you learn
- Plan your study time
- Take notes and use your notes to study from; your notes can be a map or an outline or any form that is most helpful for you
- Make a check-list of the areas that give you trouble and refer to this list so you practice what is difficult for you
- Problem solve
  discover what the problem (question) is
  list two or three possible solutions
  choose the one best answer
  try it out
  re-think and research if it does not work.

Try to find out why a solution is wrong and keep that in mind to apply to future material you read.

The on-going repetition and review of both the test-taking and study skill strategies and the constant practice will give you confidence and self-assurance as you prepare for and take the GED exam.

PRETEST
Science

DIRECTIONS: Choose the one best answer for each item below.

Items 1–4 refer to the following passage and graph.

Animals that do not have backbones or backbone-like structures are classified as invertebrates. Animals with backbones or backbone-like structures are classified as vertebrates. Each of these main classifications is divided into smaller groups based on common characteristics. These groups are called phyla. The pie graph below shows the relative size of various animal phyla.

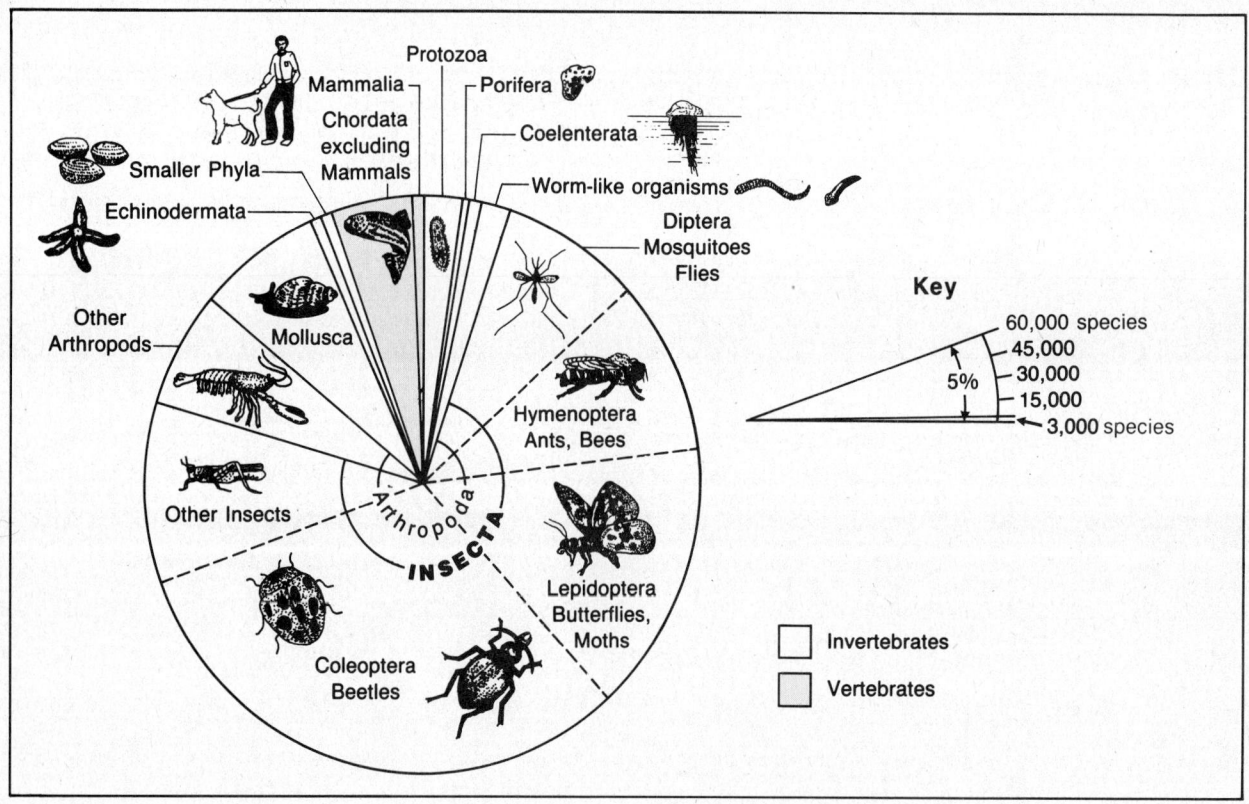

1. According to the graph, approximately what percent of all animals are invertebrates?

   (1) 100%
   (2) 95%
   (3) 75%
   (4) 50%
   (5) 5%

2. Mammalia is the name of a class of animals belonging to the phylum Chordata. According to the graph, approximately how many species of mammals are there?

   (1) 65,000
   (2) 50,000
   (3) 15,000
   (4) 10,500
   (5) 3,500

GO ON TO THE NEXT PAGE.

3. Insecta names the class of animals to which insects belong. The class Insecta is divided into several orders. According to the graph, which of the following orders is part of that class?

   (1) Chordata
   (2) Coelenterata
   (3) Lepidoptera
   (4) Mammalia
   (5) Protozoa

4. Arthropoda is the name of a phylum that includes insects. According to the graph, approximately how many species of arthropoda are there, excluding insects?

   (1) 600,000
   (2) 60,000
   (3) 30,000
   (4) 10,000
   (5) 6,000

Items 5–8 refer to the following passage and diagram.

Leaves are complex structures made up of several different kinds of tissues and cells. The top and bottom of a leaf are covered by a thin, protective layer of cells called the epidermis. The epidermis is covered by a waxy substance called the cuticle. The cuticle helps to waterproof the leaf. Between the upper and lower epidermal layers are cells that contain chloroplasts. Chloroplasts are where the plant makes its food through the process of photosynthesis. Running throughout the leaf are fibrovascular bundles. These bundles appear as veins when we look at a leaf. They conduct water up to the cells of the leaf and carry the food manufactured by the leaf down to other parts of the plant.

On the underside of the leaf are many openings called stoma. Each stomata is surrounded by specialized guard cells. The guard cells regulate the opening and closing of the stoma. By so doing, they control the amount of air and, therefore, carbon dioxide that enters the leaf and the amount of oxygen and water that leaves it.

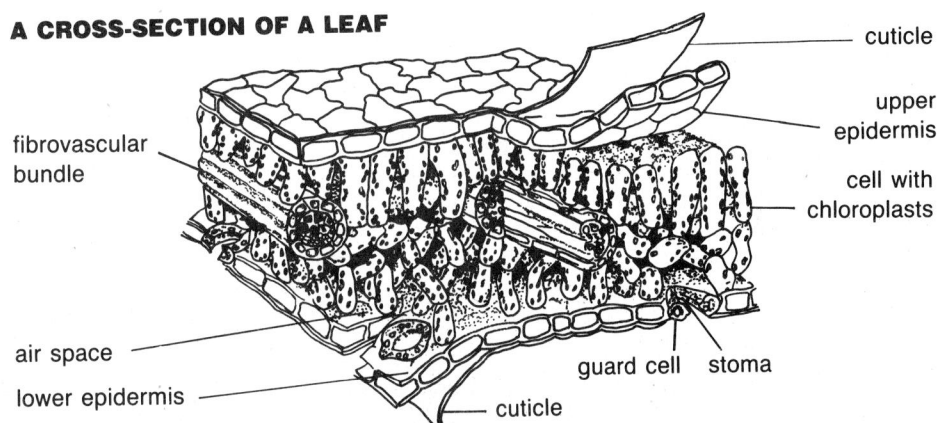

A CROSS-SECTION OF A LEAF

5. When water hits a leaf, it usually beads up and rolls off. Which of the following leaf structures is responsible for this?

   (1) epidermis
   (2) chloroplasts
   (3) cuticle
   (4) fibrovascular bundles
   (5) stoma

6. Which of the following would likely happen if there were no guard cells to control the amount of water leaving the plant?

   (1) The plant would wilt from lack of water.
   (2) The cuticle would wear away.
   (3) The epidermis would rupture.
   (4) The leaves would turn toward the sun.
   (5) The stoma would close.

GO ON TO THE NEXT PAGE.

7. Most of the water a plant needs enters through the root system. Which leaf structure distributes this water to the leaf's cells?

(1) epidermis
(2) stoma
(3) guard cells
(4) cuticle
(5) fibrovascular bundles

8. Chloroplasts use the sun's energy to make food. How does this explain the fact that most cells containing chloroplasts are located near the upper surface of a leaf?

(1) The upper surface of the leaf receives the most sunlight.
(2) The upper surface of the leaf receives the least sunlight.
(3) The lower surface of the leaf receives the most sunlight.
(4) The upper surface of the leaf has a thicker epidermis.
(5) Sunlight bounces off the upper surface of the leaf.

Items 9–12 refer to the following passage and diagram.

Unlike flowering plants, mosses do not produce seeds. Instead they produce spores that grow into separate plants that produce male and female branches. The leafy part of a moss plant is called a gametophyte. That is the part of the plant most often seen. A stalk with a capsule on the end grows out of the gametophyte. The stalk with its capsule is called a sporophyte. The sporophyte produces spores that are released into the air. If they land where conditions are right, a new gametophyte begins to grow. As the young gametophyte matures, it develops a male structure called an antheridium, which produces sperm. Other branches develop a female structure called an archegonium, which produces eggs. The sperm move down the male branch, up the female branch and into the egg case where they fertilize some of the eggs. The fertilized eggs are called zygotes. The zygotes soon begin to develop into a new sporophyte and the entire process begins again.

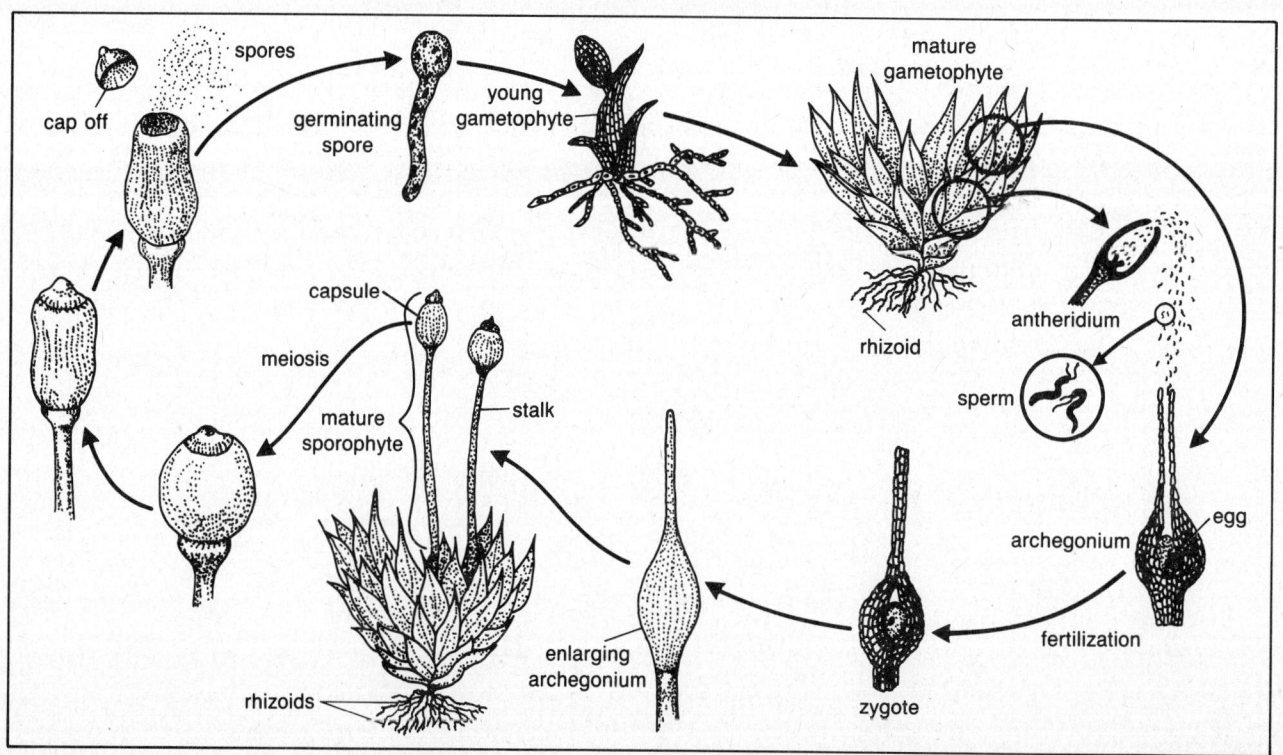

GO ON TO THE NEXT PAGE.

9. According to the diagram, the structure from which spores are released is called

(1) a gametophyte
(2) a rhizoid
(3) an archegonium
(4) an antheridium
(5) a capsule

10. A germinating spore develops into

(1) a sporophyte
(2) a young gametophyte
(3) an antheridium
(4) a zygote
(5) an egg

11. The zygote develops in the archegonium, which is on the

(1) sporophyte
(2) antheridium
(3) stalk
(4) capsule
(5) gametophyte

12. Which of the following would be the *best* title for the diagram?

(1) The Life Cycle of a Moss
(2) The Development of Spores
(3) The Difference between Flowering Plants and Mosses
(4) How Plants Grow
(5) The Evolution of Mosses

Items 13–14 refer to the following passage.

A solution is a homogeneous mixture in which one substance, called the solute, is dissolved in another substance, called the solvent. Although solutions can be made from solids, liquids or gases, many common solutions have a solid solute and a liquid solvent.

The amount of solute that can be dissolved in a given amount of solvent is called solubility. Solubility is dependent upon temperature. For a solid dissolved in a liquid, solubility increases as temperature increases.

How quickly a solute dissolves in a given solvent is called the rate of solution. For solids dissolved in liquids, the rate of solution increases when the energy of the solute particles increases. This happens when the solution is heated or stirred. Increasing the surface area of the solute also increases the rate of solution. Thus, a sugar cube dissolves faster when it is crushed than when it is left whole.

13. Which of the following statements can be inferred from the passage?

(1) More solute will dissolve in hot water than in other liquids of the same temperature.
(2) Heating a solution makes the particles of the solute move faster.
(3) Heating a solution makes the solute particles decrease in size.
(4) Stirring a solution increases the solubility of the solute.
(5) Equal amounts of salt and sugar will dissolve in water at room temperature.

14. A food blender increases the rate of solution because it

(1) mixes liquids faster than solids
(2) causes the heat of solution to increase
(3) increases the amount of solute that can dissolve in a given solvent
(4) chops the solute into small particles and agitates the solution
(5) increases the solubility of the solvent

GO ON TO THE NEXT PAGE.

Items 15–16 refer to the following passage.

An index fossil is a fossil that can be used to determine the relative age of the rock in which it is found. Fossils of organisms that lived for a limited period of time in many different places are used as index fossils. The best index fossils are formed from organisms that have distinctive features that make them easy to identify.

The trilobite is an example of an index fossil. The trilobite was a small, shelled sea animal that lived during the Cambrian Period. The same kinds of trilobites have been found both in the Grand Canyon and in Wales. Thus, scientists have concluded that rocks found in these areas were formed at the same time.

15. It can be inferred from the passage that the "relative age" of a rock refers to

(1) the age of the rock compared to the age of a fossil found in the rock
(2) the age of the rock compared to the age of other rock
(3) the age of the rock compared to the age of a certain organism
(4) the age of the rock compared to the age of the earth
(5) the length of time it took for the rock to form

16. According to the information provided, fossils of clams would *not* be useful as index fossils because

(1) clams are still in existence today
(2) clams have shells
(3) clams are too small to be used as index fossils
(4) clams are found in many parts of the world
(5) clams live only in the ocean

Items 17–18 refer to the following paragraph.

In an electric circuit, power, voltage and current are related by the equation, power = voltage × current. Power is measured in watts, voltage is measured in volts and current is measured in amps.

17. Which of the circuits described below could power three appliances if each appliance requires 200 watts?
  I. A circuit carrying 1.5 amps and driven by 500 volts
  II. A circuit carrying 1.0 amp and driven by 200 volts
  III. A circuit carrying 2.0 amps and driven by 300 volts

(1) II only
(2) III only
(3) I and II
(4) I and III
(5) I, II and III

18. A desk lamp can safely carry a maximum current of 0.5 amps. If the lamp operates at 150 volts, which of the following is the *smallest* bulb that would prove hazardous if used in the lamp?

(1) 40 watt
(2) 60 watt
(3) 75 watt
(4) 100 watt
(5) 150 watt

GO ON TO THE NEXT PAGE.

Items 19–20 refer to the following passage.

Compounds that contain carbon, such as petroleum, are known as organic compounds. Because carbon-containing compounds are present in all living things, it was once believed that organic compounds could not be synthesized in the laboratory. Then in 1828 the German scientist Friedrich Wohler synthesized the organic substance urea. From this time on, scientists recognized that organic compounds could be produced from inorganic materials. Today the majority of organic compounds are synthesized in chemical laboratories.

Over 90% of all known compounds are organic. One of the most important sources of organic compounds is petroleum. From petroleum, fuels such as gasoline, kerosene and diesel are produced. Petroleum is also the main source of organic compounds used to produce such products as plastics, synthetic fabrics, paint and detergent.

19. Wohler's work was significant because he

(1) determined the structure and composition of urea
(2) discovered new uses for petroleum
(3) proved that organic compounds are present in all living things
(4) produced the first synthetic organic compound
(5) proved that urea was an organic compound

20. Which of the following statements is supported by the passage?

(1) Fewer organic compounds are present in living things today than were present several hundred years ago.
(2) Only a small percentage of living things contain organic compounds.
(3) Without petroleum it would be difficult to produce many of the products we use every day.
(4) Only a small percentage of organic compounds are useful in industry.
(5) Most organic compounds come from petroleum.

Items 21–22 refer to the following information.

The rule for magnetic poles is: Opposite poles attract each other while like poles repel each other. The size of the force of attraction or repulsion between two magnetic poles will be directly proportional to the product of the strengths of the poles and inversely proportional to the square of the distance between the poles.

21. North and south magnetic poles of equal strength are separated by a distance of 20 centimeters. If the north pole is moved closer to the south pole, which of the following events will occur?

(1) The force of repulsion between the two poles will increase.
(2) The force of repulsion between the two poles will decrease.
(3) The force of attraction between the two poles will increase.
(4) The force of attraction between the two poles will decrease.
(5) The forces between the two poles will remain the same.

22. Two magnetic poles of equal strength are separated by a distance of 2 centimeters. One pole is replaced by a pole four times as strong. If the size of the force between the two poles is to remain the same, then

(1) the distance between the two poles must remain the same
(2) the distance between the poles must be increased to 8 centimeters
(3) the distance between the two poles must be reduced to 1 centimeter
(4) the distance between the two poles must be increased to 4 centimeters
(5) the distance between the poles must be increased to 3 centimeters.

Answers and Explanations

Science Pretest pp. 4–9

1. **Answer:** (2) By using the scale that accompanies the graph, it can be determined that about 95% of all animals are invertebrates. The quickest way to figure this out is to first compare the size of the vertebrate group with the scale. Then subtract that 5% from 100% to arrive at the size of the invertebrate group.

2. **Answer:** (5) Again using the scale that accompanies the graph, it can be determined that the class Mammalia, to which mammals belong, is made up of about 3,500 species. None of the other choices is responsible, according to the scale.

3. **Answer:** (3) To answer the question correctly you must understand which part of the graph represents the class Insecta. That can be determined by identifying the inner circle labeled "Insecta." Of the orders included in this class, only Lepidoptera is listed as a choice.

4. **Answer:** (2) The label on the graph you are looking for is "Other Arthropods." Once you have identified that group, you can use the scale to determine about how many species there are.

5. **Answer:** (3) According to the passage, the cuticle makes the leaf waterproof. It is because the leaf is waterproof that water beads and rolls off it. Each of the other choices names structures that have entirely different functions from that of the cuticle. The epidermis, choice (1), is covered by the cuticle so would be unaffected by water striking the surface of the leaf.

6. **Answer:** (1) Most likely too much water would escape from the leaf, causing the plant to wilt. The guard cells have nothing to do with choices (2), (3) and (4). The stoma, choice (5), could not close without guard cells.

7. **Answer:** (5) According to the article, the fibrovascular bundles carry water to the cells of the leaf. This fact is not changed even though the water enters the plant through the root system.

8. **Answer:** (1) If the chloroplasts need the sun's light to make food, it stands to reason that the cells containing them would be located where they would receive the most sunlight. The upper surface of a leaf is turned toward the sunlight; therefore, you would expect to see cells containing chloroplasts near the upper surface of the leaf.

9. **Answer:** (5) The diagram clearly labels the capsule as the structure from which the spores are released. Choice (1) is not part of the sporophyte. Choice (2) is what anchors the moss to the ground. Choices (3) and (4) are the names of the female and male structures of the moss plant.

10. **Answer:** (2) According to the diagram, a germinating spore develops into a young gametophyte. Choice (1) develops from the zygote, choice (4). Choice (3) is the name of the male structure on the gametophyte. Choice (5) is the name of the female sex cell.

11. **Answer:** (5) According to the passage and diagram, the archegonium is the female structure that develops on the young gametophyte. The sperm moves to the archegonium and there fertilizes the egg to form a zygote, which develops into the sporophyte. The sporophyte, therefore, develops out of the archegonium on the gametophyte.

12. **Answer:** (1) The diagram shows the life cycle of a typical moss plant. Choice (2) is included in the diagram, but only part of it. The diagram does not represent choices (3), (4) and (5).

13. **Answer:** (2) Heating a solution increases the energy of solute particles, which would cause the particles to move faster. Choices (1) and (5) are incorrect because no information is given regarding solubilities of specific substances. Choice (3) is incorrect because stirring is related to the energy of the particles, not to their size. Choice (4) is incorrect because stirring affects the rate of solution, not of solubility.

14. **Answer:** (4) This is correct because increasing the surface area of solute particles and stirring a solution both increase the rate of solution. Choices (1), (2) and (3) could be true, but none of these factors relates to rate of solution. Choice (5) is obviously wrong because solubility is measured in terms of the solute, not the solvent.

15. **Answer:** (2) This is correct because index fossils are used to compare the age of rocks in different locations. Choice (1) is incorrect because the rock and fossil would both be the same age. Choices (3), (4) and (5) are not relevant to the information given.

16. **Answer:** (1) This is correct because, to be useful, the organisms that form index fossils must have lived for only a short period of time. Choices (2), (3) and (5) are incorrect because these are all characteristics of the trilobite. Choice (4) is incorrect because one of the criteria for an index fossil is that it be found in many different places.

17. **Answer:** (4) This is correct because the three appliances would require a minimum of 600 watts. Circuit I provides 750 watts, and circuit III provides 600 watts. Choices (1), (3) and (5) are incorrect because circuit II provides only enough power for one appliance at 200 watts. Choice (2) is incorrect because having more than the minimum 600 watts is just as adequate.

18. **Answer:** (4) This is correct because the maximum number of watts that can be supported by the lamp is 75. Choices (1), (2) and (3) are incorrect because these bulbs can all be operated safely. Choice (5) is incorrect because it is not the *smallest* bulb that would prove hazardous.

19. **Answer:** (4) This is correct because, until Wohler's synthesis of urea, scientists believed that organic compounds could not be synthesized. Choices (1), (3) and (5) are incorrect because it can be assumed from the passage that these things were known before Wohler produced urea. Choice (2) is incorrect because no mention is made of Wohler and petroleum.

20. **Answer:** (3) This is correct because the passage indicates that many familiar products are made from organic compounds that come from petroleum. Choice (1) is incorrect because there is nothing to suggest that the composition of living things would change over a few hundred years. Choice (2) is incorrect because organic compounds are present in all living things. Choice (4) is incorrect because the passage emphasizes the usefulness of organic compounds in industry. Choice (5) is incorrect because the passage relates the importance of petroleum to the usefulness, not the number, of organic compounds it provides.

21. **Answer:** (3) Opposite poles attract each other and the force of attraction increases as the distance between the two poles decreases. Choices (1) and (2) are incorrect because opposite poles do not repel each other. Choices (4) and (5) are incorrect because decreasing the distance will cause the force between them to increase.

22. **Answer:** (4) This is correct because doubling the distance is compensated by quadrupling the strength of one pole. (This works because $2^2 = 4$.) Choices (1), (3) and (5) are incorrect because they result in forces larger than the original. Choice (2) is incorrect because it results in a force smaller than the original.

To figure out your score, count the problems you missed. Then subtract the number of problems you missed from the total number of questions on the test. If half or more are correct, you may consider that you have passed the test. To organize your study time efficiently, turn to the Pretest/Posttest Diagnostic Chart in the back of this book.

ANSWERS AND EXPLANATIONS

OVERVIEW
Life Science

Belted kingfisher chick hatching from an egg.

environment
all the surrounding conditions in which an organism lives

hereditary
capable of being passed from one generation to the next

Life science is the field of science that studies living things and how they interact with each other and their **environment**. Understanding life science concepts helps us to understand ourselves and the world around us. It provides us with the opportunity to improve the quality of our life and to protect the environment in which we live.

In the first lesson of this section you will learn about the structure of plant and animal cells. Cells are the microscopic units of which all living things are made. They take in food, carrying out hundreds of chemical reactions and reproduce. You will learn how, through the process of reproduction, cells pass on **hereditary** information from one generation to the next.

In the second lesson you will learn how plants make their own food through the process of **photosynthesis,** and how all living organisms use the process of **respiration** to obtain the energy they need. Then you will see how photosynthesis and respiration interact to create the carbon-oxygen cycle. You will learn how nitrogen is cycled through the **biosphere.**

The third lesson discusses theories for the origin of life and how living things evolved from tiny, single-celled organisms to the complex plants and animals we see around us today. You will learn how the environment affects the development of new **species** through the processes of natural selection.

The fourth lesson explains how, from a single seed, plants grow to develop such structures as stems, leaves, roots and flowers. Then you are introduced to the system scientists use to classify all living organisms. You will learn how this system of classification is based on the idea that all living things are related to one another.

In the fifth lesson you will learn about some of the **organ systems** of the human body. For example, you will learn how the digestive system works and how the heart pumps blood throughout the body. You will also learn about the endocrine system and the essential chemicals that regulate many of the body's functions.

The sixth lesson begins with a discussion of how your body protects itself from disease. Then you are introduced to two types of disease-causing organisms, bacteria and viruses, and learn about their structure and how they inflict the diseases they cause. At the end of the lesson you will learn about some scientific developments used to combat **infectious** diseases.

The seventh lesson discusses how parent organisms pass on physical characteristics to their **offspring.** You will learn about a monk called Gregor Mendel who became the father of modern **genetics,** and study some of his experiments. Finally, you will see how all genetic information is duplicated at the molecular level of the cell.

In the last lesson you will learn how living organisms interact with each other and their environment in natural communities called **ecosystems.** The lesson closes with a discussion of how environmental pollution has affected all living things.

photosynthesis
the process by which plants use light energy to make food

respiration
the chemical process of releasing energy for use by living things

biosphere
the thin layer of the earth where life exists

species
a group of similar organisms that can mate and produce fertile offspring

organ system
several organs working together to perform a specific function

infectious
capable of being spread from one organism to another

offspring
the direct descendants of an animal or plant

genetics
the study of inherited characteristics

ecosystem
a selected area where living and nonliving things interact

1 The Biology of Cells

All living organisms are made up of microscopic units called **cells.** Some organisms consist of only a single cell. Other organisms are made up of many cells. Almost all cells have certain things in common, and their basic structure is similar. Most cells take in food, metabolize it and give off its waste. Cells also grow, reproduce and die.

Most cells have a recognizable structure called a **nucleus,** which is surrounded by a nuclear membrane. The nucleus controls the cell's activities. Within the nucleus is a **nucleolus,** which plays a role in making protein. Throughout the nucleoplasm are thin, dark strands of chromatin. During cell division, the chromatin contracts to form chromosomes, which carry the hereditary information for the cell.

The cytoplasm of a cell is said to be all the material inside the cell membrane except the nucleus. The cytoplasm contains structures (organelles) that include the mitochrondria, endoplasmic reticulum, ribosomes, lysosomes, Golgi apparatus and vacuoles.

Mitochondria appear as rod-shaped organelles floating in the cytoplasm. A cell may contain hundreds of mitochondria, which produce almost all the energy the cell needs for its activities.

Lysosomes are smaller than mitochondria. Lysosomes contain enzymes for breaking down proteins. When a cell takes in large molecules, lysosomes break them down into smaller molecules that the cell can use. When a cell dies, the lysosomes release enzymes to break down the cell's proteins. The **endoplasmic reticulum** (ER) is a system of tubes running throughout the cytoplasm. It is thought that the ER carries needed materials throughout the cell. Tiny, grain-like structures called **ribosomes** are attached to the ER and play a role in making proteins. The **Golgi apparatus** looks like stacks of flat platelets. It receives protein molecules from the ER and transports them to the cell's surface to be secreted for use in other parts of the organism. **Vacuoles** are round, open spaces in the cytoplasm. They may contain food or water for the cell's activities, or they may be used to collect and excrete waste.

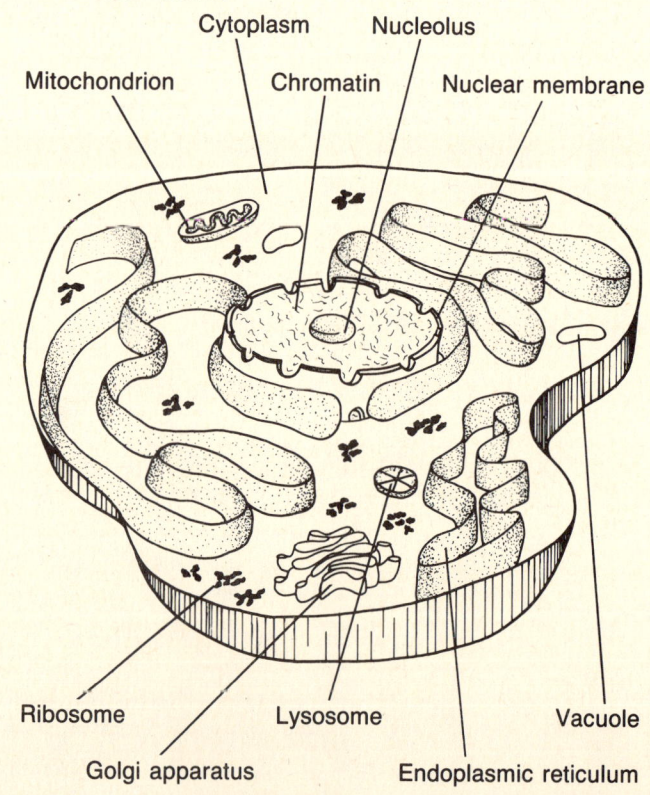

Strategies for READING

Identify the Main Idea

The main idea is the most important point an author is trying to make. It is stated at the beginning, middle or end of a paragraph.

When you read a paragraph for the **main idea,** you should focus on the first and last sentences. If either sentence seems to "sum up" the paragraph, then it is an **explicit main idea.**

Sometimes the main idea may be found in the middle of a paragraph. You can look for such a sentence if the first and the last sentence seem to give details. If there is no single sentence anywhere in the paragraph that "sums up" the ideas presented, then the paragraph does not have an explicit main idea. You will have to "read between the lines" to infer the main idea. This is called an **implicit main idea.**

When you think you have found the main idea of a passage, identify supporting details that substantiate it. If you cannot find any supporting details, then rethink the main idea.

Examples

DIRECTIONS: Use the information on this and the preceding page to choose the one best answer for each item below.

1. Which of the following sums up the main idea of the paragraph?

   (1) All cells need energy to live.
   (2) Most cells have things in common and have a similar structure.
   (3) Mitochondria give the cell energy.
   (4) The nucleus is the center of a cell.
   (5) The nucleus is an organelle.

   Answer: (2) The paragraph on the previous page states in the third sentence that "almost all cells have certain things in common and their basic structure is similar" (main idea). Choices (1) and (3) are true but do not state the main idea. Choices (4) and (5) name a structure that most cells have in common.

2. Which detail *most* completely supports the main idea?

   (1) All cells need energy to live.
   (2) Most cells have things in common and have a similar structure.
   (3) Mitochondria give the cell energy.
   (4) The nucleus is the center of a cell.
   (5) The nucleus is an organelle.

   Answer: (1) Choices (3), (4) and (5) are structures that cells have in common, so each in part supports the main idea. But, since the main idea is that cells have things in common and have a similar structure, the one choice that most completely supports the main idea is that all cells need energy to live.

Practice

When reading test questions, it is helpful to identify the main idea of the question and think about details that support it.

DIRECTIONS: Choose the one best answer for each item below.

Items 1–6 refer to the following passage and diagram.

The process by which a cell divides is called **mitosis.** During mitosis, the chromosomes in the original cell, or **parent cell,** duplicate and divide into two identical sets. The process of cell division (mitosis) is divided into five phases:

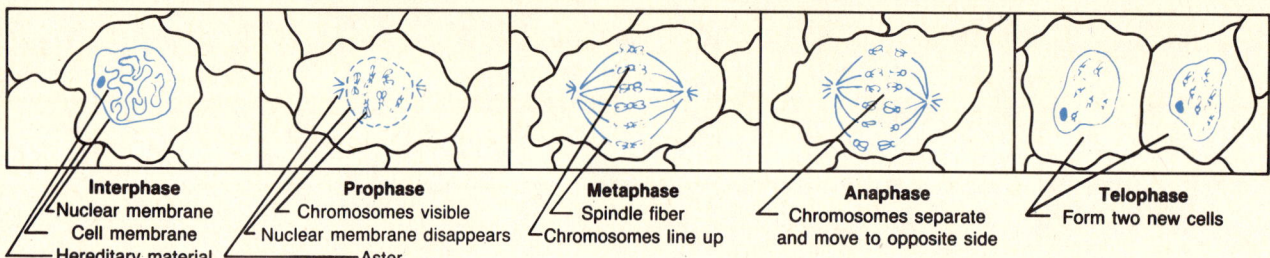

During the first phase, called **interphase,** the hereditary material in the nucleus duplicates itself. During the second phase, called **prophase,** the hereditary material shortens and thickens to form the **chromosomes.** Each chromosome is made of two identical parts that are attached to each other at their centers. Protein fibers radiate from opposite poles in the cell to form a structure called a **spindle.** The membrane surrounding the nucleus disappears. During the third phase, or **metaphase,** the chromosomes line up across the middle of the cell. The spindle fibers attach to each chromosome. During the fourth phase, called **anaphase,** the pairs of chromosomes separate, and move along the shortening spindle fibers towards opposite sides of the cell. During the final phase, or **telophase,** the chromosomes again become threads of hereditary material and the spindle fibers disappear. The membrane around the nucleus reappears, and the cytoplasm of the cell divides, producing two **daughter cells.**

1. Which phase guarantees the maintenance of identical characteristics before cell division?

   (1) interphase
   (2) prophase
   (3) metaphase
   (4) anaphase
   (5) telophase

2. According to the diagram, what purpose do the spindle fibers seem to perform in mitosis?

   (1) guide the chromosomes
   (2) give rise to the nucleus
   (3) form the nuclear membrane
   (4) form the hereditary material
   (5) form a cell plate

GO ON TO THE NEXT PAGE.

3. If the number of chromosomes in the parent cell is 48, how many chromosomes will there be in each daughter cell?

   (1) 12
   (2) 18
   (3) 24
   (4) 48
   (5) 60

4. Where is the *most* likely place for components of the nucleoli and nuclear membrane to go during prophase?

   (1) cytoplasm
   (2) nucleoplasm
   (3) chromosomes
   (4) spindles
   (5) centromeres

5. What appears to be pulling the chromatids apart during anaphase?

   (1) centriole
   (2) centromere
   (3) chromatin
   (4) nucleus
   (5) spindle

6. If mitosis were repeated three times for each daughter cell, how many cells would result?

   (1) 2
   (2) 6
   (3) 8
   (4) 16
   (5) 32

Items 7–8 refer to the following passage.

Sexual reproductive cells are formed by a process called **meiosis.** This process differs from mitosis in that the resulting cells, called **gametes,** have half the number of chromosomes as the parent cell. During sexual reproduction, two gametes combine to form a new cell called a **zygote.** Because the zygote contains the chromosomes from both gametes, the original number of chromosomes is restored.

Chromosomes occur in identical pairs. A parent cell is said to have a **diploid number** of chromosomes because it contains a complete set of paired chromosomes. A gamete receives only one chromosome from each pair in the parent cell. The resulting number of chromosomes is called a **haploid number** because it is half the number of chromosomes found in the parent cell.

7. The diploid chromosome number for human body cells is 46. What is the haploid number for humans?

   (1) 12
   (2) 23
   (3) 32
   (4) 46
   (5) 48

8. How does the chromosome number of a daughter cell resulting from mitosis differ from the chromosome number of a gamete?

   (1) It has half the number.
   (2) It has no chromosomes.
   (3) It has the same number.
   (4) It has one-fourth the number.
   (5) It has twice the number.

GED Mini-Test 1

TIP

As you take a GED Mini-Test, relax. Remember it is designed to help you build your confidence in test taking. Set your goals for the test. Then use the Answers and Explanations section to help you analyze how you did.

DIRECTIONS: Choose the one best answer for each item below.

Items 1–4 refer to the following passage.

Plant cells have some special structures that animal cells do not have. For example, most plant cells have a cell wall surrounding the membrane. A cell wall is composed of several layers. The middle layer, called the middle lamella, contains the jelly-like substance pectin. Primary walls on either side of the middle lamella are made of cellulose and pectin. The thicker the primary walls, the more rigid are the parts of the plant these cells make up. Secondary walls form on the outside of the primary walls. These rigid walls are made of cellulose and remain long after the cells have died.

Plant cells contain organelles called plastids. Some plastids store food while others manufacture it. Chloroplasts are one kind of plastid. Chloroplasts contain the green pigment chlorophyll. During the process of photosynthesis, chlorophyll uses solar energy to produce glucose and oxygen from water and carbon dioxide. Glucose is used by the plant for food. Another kind of plastid is called a leucoplast. Leucoplasts contain enzymes that convert glucose molecules into starch molecules. The starch molecules are then stored in the leucoplasts.

1. Which of the following structures must a plant contain for the manufacture of food?

    (1) glucose
    (2) nucleus
    (3) cell wall
    (4) chloroplast
    (5) leucoplast

2. Which of the following plant structures is likely to be made up of thick primary walls?

    (1) flowers
    (2) soft fruits
    (3) stems
    (4) soft vegetables
    (5) buds

3. For which of the following is the process of photosynthesis responsible?

    (1) cell division
    (2) transport of water
    (3) absorption of nitrogen
    (4) rigidity of stems
    (5) production of glucose

4. Which cellular structure is *most* responsible for wood retaining its rigidity after a tree has died?

    (1) the secondary cell wall
    (2) the middle lamella
    (3) the leucoplasts
    (4) the primary cell wall
    (5) the plastids

GO ON TO THE NEXT PAGE.

Items 5–6 refer to the following passage and diagram.

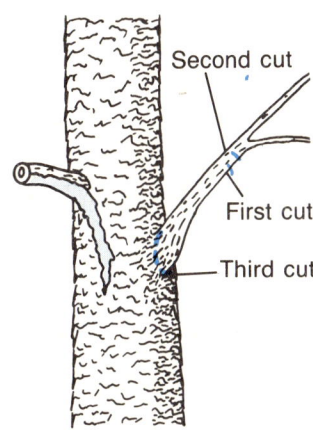

Like all living things, trees are subject to disease, decay and death. When a tree is wounded, fungus spores lodge in the wound, germinate and send out creeping threads that attack the cell tissues. In time the tree dies unless a tree surgeon saves it.

Injured or rotted branches must be removed close to the trunk, or parent branch, so as not to leave the projecting stub. When a large branch is sawed off, it should first be undercut to prevent stripping and tearing the bark as the limb falls. The first cut is made from the underside about 12 to 18 inches from the trunk and continues until the saw begins to bind from the pressure of the limb. The next cut is from above and about two inches outward from the first cut. The limb will fall without tearing away the bark and wood from the underside of the limb and the trunk. Then the stub is removed near the trunk and the wound is treated.

5. In the figure above, the limb on the left was not properly cut. What was done wrong?

   (1) The bark was stripped.
   (2) The limb was not undercut or cut close to the trunk.
   (3) Fungus threads killed it.
   (4) The rotted wood was sterilized.
   (5) The cambium grew too much.

6. What was the result of the improper cut?

   (1) The third cut was not necessary.
   (2) The natural swaying of the tree was prevented.
   (3) The bark was stripped and torn.
   (4) The bark grows inward and heals.
   (5) Fungi continued to grow.

Check your answers to the GED Mini-Test on page 20.

Answers and Explanations

Practice pp. 16–17

1. **Answer:** (1) According to the article, the hereditary material that makes up the chromosomes is duplicated during the first phase, or interphase. The already duplicated chromosomes are present during each of the other phases, choices (2) through (5).

2. **Answer:** (1) By observing the diagrams and reading the article, it appears as though the spindle fibers guide the chromosomes towards opposite sides of the cell. They could not give rise to the nucleus, choice (2), or form the nuclear membrane, choice (3), since they have disappeared already. The hereditary material, choice (4), has already become chromosomes. There is no cell plate discussed, choice (5).

3. **Answer:** (4) According to the article, each daughter cell will have the same number of chromosomes as the parent cell. Since the parent cell has 48 chromosomes, each daughter will also have 48 chromosomes.

4. **Answer:** (1) Since the components of the nucleoli and nuclear membrane are contained within the cell, it stands to reason that when they break down the only place their component parts can go is into the cytoplasm. Once the nuclear membrane is no longer present, the cytoplasm and what is left of the nucleus will intermix. Choice (2) is not relevant because there is no nucleoplasm by this time. Choices (3) through (5) are all structures in and of themselves.

5. **Answer:** (5) By observing the diagrams, it is apparent that the spindle fibers are pulling the chromatids apart. Choice (1) remains in one place and does not change its appearance. Choice (2) has no apparent means of pulling the two chromatids apart. Choices (3) and (4) are not relevant in anaphase.

6. **Answer:** (4) Since each daughter cell will become a parent cell and divide into two more cells, three divisions will result in 16 new cells.

7. **Answer:** (2) According to the article, the haploid number is half the diploid number. Since the human diploid number is 46, half of it, or the haploid number, is 23.

8. **Answer:** (5) A gamete is the result of meiosis. Since the number of chromosomes resulting from meiosis is half that of a daughter cell resulting from mitosis, the daughter cell will have twice the number of chromosomes as the gamete.

GED Mini-Test pp. 18–19

1. **Answer:** (4) Photosynthesis occurs in the chloroplast. Choice (1) is the food manufactured in the chloroplast. Choice (2) is the part of the cell that controls the cell's activities. Choice (3) provides rigidity to the cell, and choice (5) stores the glucose after it is manufactured by photosynthesis in the chloroplast.

2. **Answer:** (3) According to the passage, the thicker the primary walls of the cell wall, the more rigid are the parts of the plant these walls make up. Since stems are the most rigid structure listed, choice (3) is the correct answer. The word "soft" used to describe choices (2) and (4) is a clue that these two choices are incorrect. And flowers and buds are also soft, so they are incorrect.

3. **Answer:** (5) According to the passage, the products of photosynthesis are glucose and oxygen. Choice (1) is controlled by the nucleus of the cell. Choices (2), (3) and (4) are not relevant to photosynthesis.

4. **Answer:** (1) Since the secondary cell wall is made rigid by cellulose and often remains after the cell has died, choice (1) is the only correct choice. The middle lamella, choice (2), is made up of pectin, a jelly-like substance, which would not make wood rigid. Choices (3) and (4) are organelles that have nothing to do with cell rigidity. The primary cell wall, choice (4), is made up of cellulose and pectin and would not be rigid enough to maintain wood's rigidity, especially after the tree has died.

5. **Answer:** (2) As you can see from the diagram, the first undercut was not made and the limb was not cut close to the trunk, so choice (2) is the correct answer. Choice (1) was the result of the improper cut. Choices (3), (4) and (5) do not relate to this.

6. **Answer:** (3) As a result of the improper cut, the bark was stripped and torn so choice (3) is the only possible answer. The other answers really do not pertain to the improper cutting of the tree.

Photosynthesis

In one way or another, almost all animal life is dependent upon plants for food. Plants, on the other hand, produce their own food through the process of photosynthesis. Without photosynthesis, then, life as we know it would cease to exist.

Photosynthesis is the process by which green plants, using chlorophyll and the energy of sunlight, produce **carbohydrate** (glucose) out of water and carbon dioxide, liberating oxygen in the process.

During photosynthesis, plants capture the sun's energy and use it to make **glucose** ($C_6H_{12}O_6$), a form of carbohydrate. Glucose is used to build other substances that the plant needs, such as starches and proteins. Oxygen (O_2) is a **by-product**, or waste material, of photosynthesis. Without this supply of oxygen, the oxygen in the atmosphere would quickly decrease.

Photosynthesis is divided into two phases—the light reactions and the dark reactions. The **light reactions** must take place in the presence of sunlight. Light energy from the sun is trapped by the plant's **chlorophyll,** the green pigment. It is changed by the chlorophyll molecules into chemical energy and in turn is used to split water molecules (H_2O) into hydrogen and oxygen. The **dark reactions** are so named because light is not required for them to take place. During this phase of photosynthesis, the hydrogen produced in the light reaction combines with carbon dioxide (CO_2) from the atmosphere to form glucose. This chemical reaction takes place in the presence of **enzymes,** chemicals that are necessary for certain reactions to occur. The rest of the hydrogen and oxygen combines to form water. These two chemical reactions of photosynthesis occur in a fraction of a second.

The chemical equation for photosynthesis combines the reactions that occur in both phases and may be expressed as follows:

$$12H_2O + 6CO_2 \xrightarrow[\text{chlorophyll}]{\text{light}} C_6H_{12}O_6 + 6O_2 + 6H_2O$$

water + carbon dioxide → glucose + oxygen + water

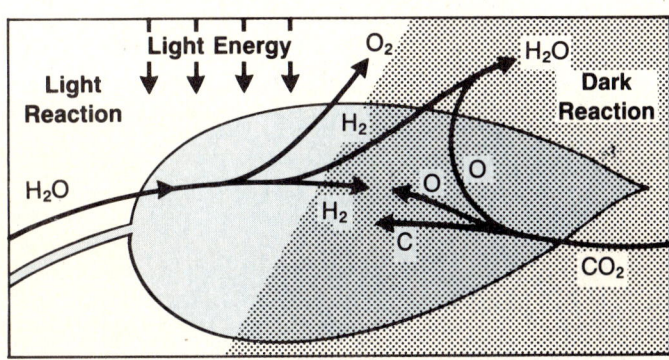

INTRODUCTION 21

Strategies for READING

Restate Information

This skill involves identifying the main idea, topic sentence and supporting details in order to present information in another way. It can often help you better understand something you are reading or studying.

There are many different ways to **restate information.** A chart, a graph, a diagram, a main idea sentence or an equation are some frequently used ways. Sometimes a paragraph may be written without a stated main idea. You can determine the main idea by deciding what point the supporting details make and then restating the details in a topic sentence.

Read the details in the second paragraph on page 21. Try to write a topic sentence. If you wrote a main idea sentence such as *"The process of photosynthesis produces glucose and oxygen,"* you understood that the details were discussing the outcome of the process of photosynthesis.

In the article, photosynthesis is stated in three different ways. One way is a written explanation. What are the other two ways? (a diagram and a chemical equation) Which way is easiest for you to understand? Do all three ways help you understand the information better?

Examples

DIRECTIONS: Use the information on this and the preceding page to choose the one best answer for each item below.

1. Which element produced in the light reaction of photosynthesis combines with CO_2 to produce glucose in the dark reaction?

    (1) oxygen
    (2) carbon
    (3) hydrogen
    (4) chlorophyll
    (5) water

Answer: (3) Hydrogen formed in the light reaction combines with CO_2 to form glucose. Choices (4) and (5) are not elements (restatement).

2. During the process of photosynthesis, which atoms combine to form a molecule of glucose?

    (1) hydrogen and oxygen
    (2) carbon and oxygen
    (3) carbon and hydrogen
    (4) carbon, hydrogen and oxygen
    (5) carbon, hydrogen and nitrogen

Answer: (4) The chemical symbol for glucose, $C_6H_{12}O_6$, can be *restated* as 6 atoms of carbon, 12 atoms of hydrogen and 6 atoms of oxygen. Nitrogen, choice (5), is not involved in the process of photosynthesis.

Practice

If some of the information you are reading or studying is not clear, think of ways to restate it.

DIRECTIONS: Choose the <u>one</u> best answer for each item below.

Items 1–4 refer to the following passage.

Respiration is the process by which cells obtain the energy they need in order to function. During respiration, glucose molecules are *broken down* to release chemical energy. As a result, carbon dioxide and water are released. Many of the reactions in respiration are the opposite of those in photosynthesis where glucose is *formed* from water and carbon dioxide.

There are two kinds of respiration—direct and indirect respiration. **Direct respiration** occurs in single-celled organisms, like the amoeba, that are able to directly exchange gases with the environment through their membranes. **Indirect respiration** occurs in multicellular organisms where most of the body's cells are not in direct contact with the environment. Indirect respiration can be divided into two phases. **External respiration** exchanges gases between the environment and the blood. Gills and lungs perform this function. **Internal respiration** exchanges gases between the blood and cells. The circulatory system provides this function. The chemical equation for respiration is:

$$C_6H_{12}O_6 + 6O_2 \rightarrow 6CO_2 + 6H_2O + energy$$
glucose   oxygen   carbon dioxide   water

1. Which of the following organisms functions with direct respiration?

    (1) human
    (2) bird
    (3) lizard
    (4) amoeba
    (5) dog

2. Which two substances are the direct result of respiration?

    (1) oxygen and hydrogen
    (2) carbon dioxide and water
    (3) glucose and water
    (4) glucose and hydrogen
    (5) water and oxygen

3. Which two substances would you expect to be exhaled by the lungs?

    (1) oxygen and carbon dioxide
    (2) hydrogen and oxygen
    (3) glucose and carbon dioxide
    (4) hydrogen and carbon
    (5) water vapor and carbon dioxide

4. If an animal's lungs were damaged, what would be the *most* likely effect on respiration? Respiration would

    (1) be speeded up
    (2) be slowed down
    (3) not change
    (4) stop
    (5) become direct

GO ON TO THE NEXT PAGE.

PRACTICE 23

Items 5–8 refer to the following diagram and passage.

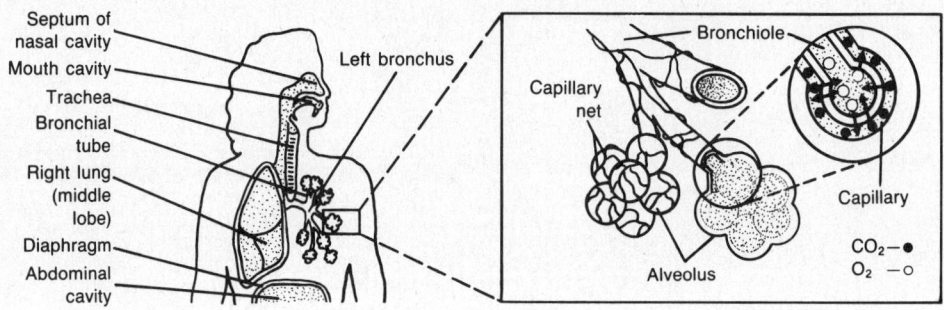

When we breathe, air is taken in through the nose or mouth and passes down through the **trachea,** which extends from the back of the mouth to the lungs. The trachea then divides into two **bronchi,** and each **bronchus** divides into several bronchial tubes. Each **bronchial tube** divides into still smaller tubes called **bronchioles.** And the bronchioles end in small air sacs called **alveoli** (the singular is **alveolus**), which make up most of the lung's tissue.

The exchange of oxygen and carbon dioxide with the blood takes place at the alveoli, which are covered by a network of tiny blood vessels called **capillaries.** Oxygen from the air in the alveoli diffuses through the alveolar and capillary membranes into the blood stream. Carbon dioxide in the blood diffuses through the capillary and alveolar membranes into the lungs. Once in the lungs, the carbon dioxide is carried out of the body as we exhale.

5. After we inhale, what is the order of passages the air passes through to reach the alveoli?

   (1) trachea, bronchial tubes, bronchi, bronchioles
   (2) trachea, bronchi, bronchial tubes, bronchioles
   (3) trachea, bronchioles, bronchial tubes, bronchi
   (4) trachea, bronchial tubes, bronchioles, bronchi
   (5) trachea, bronchi, bronchioles, bronchial tubes

6. The pulmonary artery carries deoxygenated blood into the lungs and branches into the network of capillaries that surround the alveoli. What happens to this blood before it leaves the lungs to go to the rest of the body?

   (1) It becomes more deoxygenated.
   (2) It becomes saturated with carbon dioxide.
   (3) It diffuses into the alveoli.
   (4) It receives oxygen from the alveoli.
   (5) It receives carbon dioxide from the alveoli.

7. There is a small amount of carbon dioxide in the air we breathe. What *most* likely happens to this carbon dioxide during the process of respiration?

   (1) It remains in the lungs until we exhale.
   (2) It diffuses into the capillaries.
   (3) It diffuses into body cells.
   (4) It remains in the bronchi.
   (5) It diffuses out of the alveoli.

8. Gases diffuse from an area of greater concentration to an area of lesser concentration. Which of the following *best* describes the concentration of oxygen in the alveoli with respect to the concentration of oxygen in the capillaries?

   (1) equal in both
   (2) higher in the blood
   (3) higher in the capillaries
   (4) lower in the alveoli
   (5) higher in the alveoli

Before you take the GED Mini-Test, check your answers on pages 26–27.

GED Mini-Test 2

TIP Carefully read each question twice before you select an answer. Reread the question and the answer you chose before marking your answer sheet.

DIRECTIONS: Choose the one best answer for each item below.

Items 1–4 refer to the following diagram and passage.

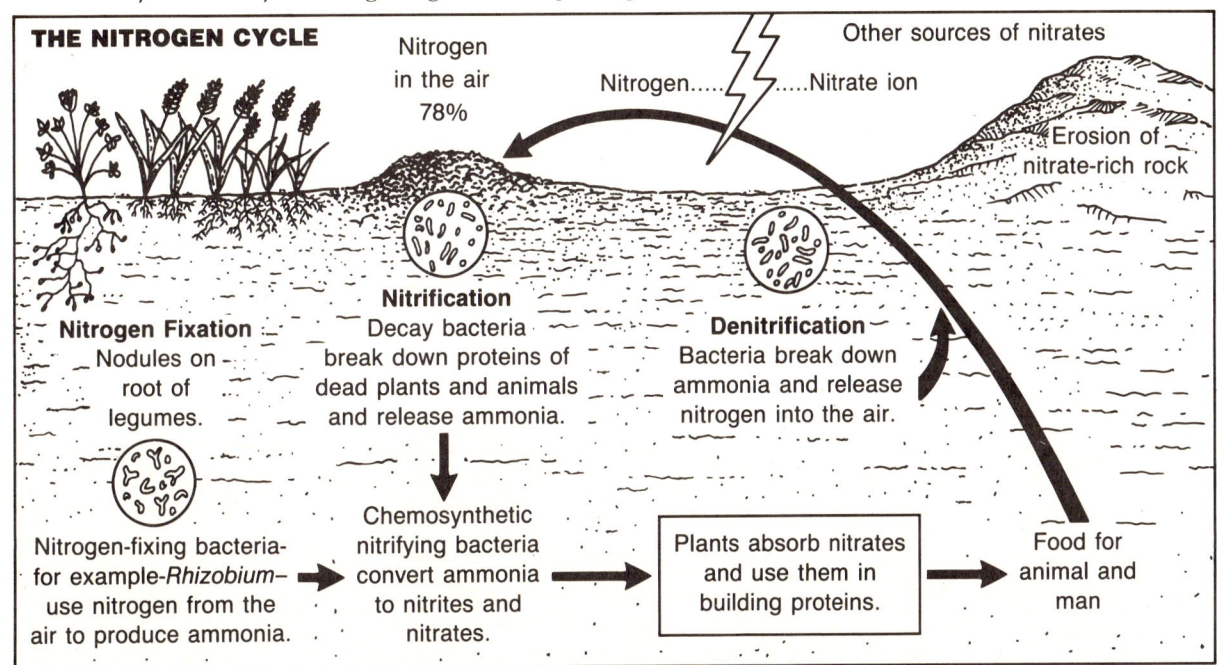

Plants and animals must have nitrogen in order to make amino acids that are in turn combined to form proteins. Animals receive the nitrogen they need by eating plants or other animals that eat plants. Plants, however, cannot use the element nitrogen directly. Instead, plants must get the nitrogen they need from nitrogen-rich compounds. Bacteria play the most important role in supplying plants with nitrogen. Through a process called nitrogen fixation, certain kinds of bacteria can change atmospheric nitrogen into compounds that plants can use. These bacteria live in the roots of certain plants called legumes. Through the process of nitrification, bacteria break down the proteins of decaying organisms into nitrate and nitrite compounds, making them available to plants. Other bacteria return nitrogen to the atmosphere by breaking down ammonia, a nitrogen compound, in the soil. This process is called denitrification.

1. What is the importance of nitrogen to plant life?

   (1) It is used by plants to make glucose.
   (2) Plants cannot make protein without it.
   (3) Animals depend on plants for it.
   (4) It combines with sulfur and phosphorus.
   (5) It returns to the soil as ammonia.

2. According to the diagram, where does the process of nitrogen fixation take place?

   (1) in decaying organisms
   (2) in the leaves of plants
   (3) in the roots of all plants
   (4) in the atmosphere
   (5) in nodules on the roots of legumes

GO ON TO THE NEXT PAGE.

3. By what process do bacteria break down the protein of decaying organisms into nitrite and nitrate compounds?

   (1) nitrogen fixation
   (2) denitrification
   (3) erosion
   (4) nitrification
   (5) ionization

4. If nitrogen were not made available to plants, which of the following is the *most* likely ultimate outcome?

   (1) All animals would die.
   (2) All plants would die.
   (3) All animals and plants would die.
   (4) Animals would make their own nitrogen.
   (5) Plants would make their own nitrogen.

Items 5–6 refer to the following passage and diagram.

The two main processes involved in the carbon-oxygen cycle are respiration and photosynthesis. Respiration takes place in almost all plants and animals. Only plants, however, are capable of photosynthesizing. During cellular respiration glucose is broken down and carbon dioxide is released. During photosynthesis water, carbon dioxide and light energy from the sun produce oxygen, glucose and water. Other sources of carbon dioxide include the decaying of dead organisms and the burning of fossil fuels.

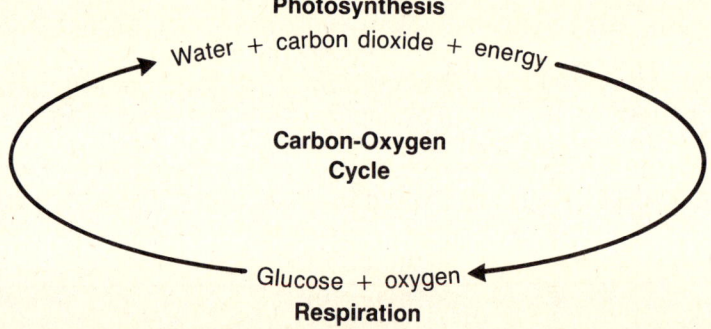

5. According to the diagram, what reacts with glucose during respiration to produce water, carbon dioxide and energy?

   (1) carbon
   (2) hydrogen
   (3) oxygen
   (4) water
   (5) carbon dioxide

6. If there were no photosynthesis, what would happen to the amount of oxygen available to animals?

   (1) It would remain unchanged.
   (2) It would increase.
   (3) It would go into the atmosphere.
   (4) It would decrease.
   (5) It would be absorbed by plants.

Check your answers to the GED Mini-Test on page 27.

Answers and Explanations

Practice *pp. 23–24*

1. **Answer:** (4) Direct respiration occurs in single-celled organisms. Indirect respiration occurs in most multicellular organisms. Choice (4) is the only single-celled organism listed. Choices (1) through (3) and choice (5) are incorrect because they all name higher, multicellular organisms.

2. **Answer:** (2) According to the equation, carbon dioxide and water are the only two substances that result from respiration. Choices (1), (3), (4) and (5) are incorrect because neither hydrogen, glucose nor oxygen is the direct result of respiration.

3. **Answer:** (5) Since water and carbon dioxide are the result of respiration, it stands to reason that carbon dioxide and water vapor would be exhaled. In choice (1) oxygen is not a result of respiration. In choice (2) neither hydrogen nor oxygen is the result of respiration. In choice (3) glucose is not a result of respiration. In choice (4) neither hydrogen nor carbon is the result of respiration.

5. **Answer:** (2) According to the passage and the diagram, the air we inhale passes through the trachea, bronchi, bronchial tubes and bronchioles before arriving at the alveoli. Choice (1) and choices (3) through (5) are incorrect because the tubes are listed in the wrong order.

7. **Answer:** (1) Since carbon dioxide does not diffuse out of the alveoli into the blood, it must remain in the lungs until we exhale. For this reason, choices (2), (3) and (5) are incorrect. Choice (4) is incorrect because carbon dioxide would travel through the bronchi with the air that was inhaled.

4. **Answer:** (2) Since the exchange of gases between the environment and the blood occurs in the lungs, any damage to the lungs would impair this exchange and slow respiration down. Choice (1) is incorrect because damaged lungs would not be able to function as well as normal lungs. Choice (3) is incorrect since any damage to the lungs, which are essential to respiration, would affect respiration. Choice (4) is incorrect because it is unlikely that damaged lungs would cease to function altogether. Choice (5) is incorrect because multicellular animals cannot carry on direct respiration.

6. **Answer:** (4) Since the blood that enters the lungs has little oxygen, it picks up oxygen at the alveoli before circulating through the body. Choices (1) and (3) are irrelevant. Choice (2) is incorrect because deoxygenated blood would already contain carbon dioxide. Choice (5) is incorrect because blood never receives carbon dioxide from the alveoli.

8. **Answer:** (5) Since oxygen passes from the alveoli into the capillaries, the concentration of the oxygen in the alveoli must be higher than the concentration of oxygen in the capillaries. For this reason, choices (1) through (4) are incorrect.

GED Mini-Test pp. 25–26

1. **Answer:** (2) Choice (1) is wrong because glucose is produced first, before nitrogen is added to make amino acids and proteins. Choice (3) is true, but does not answer the question. Choices (4) and (5) are not mentioned in the passage.

3. **Answer:** (4) Nitrification is the process by which nitrogen is made available to plants by the breakdown of proteins in decaying organisms. Choices (1) and (2) name other nitrogen-producing processes. Choices (3) and (5) are irrelevant.

5. **Answer:** (3) Oxygen reacts with glucose to produce water, carbon dioxide and energy. The other items are all incorrect because they name the wrong element or compound.

2. **Answer:** (5) According to the label on the diagram, nitrogen fixation takes place only in the nodules on the roots of those plants that are legumes.

4. **Answer:** (3) Since animals depend ultimately on plants for nitrogen, and plants, themselves, need nitrogen to survive, both would die if the supply of nitrogen were cut off. Choices (1) and (2) are incorrect because they are incomplete. Choices (4) and (5) are incorrect because neither animals nor plants can produce nitrogen.

6. **Answer:** (4) Photosynthesis provides most of the oxygen for the oxygen-carbon cycle. Without it, the supply of oxygen available to animals would decrease. This is why choices (1) through (3) are incorrect. Choice (5) is incorrect because plants do not absorb oxygen even when it is available.

Evolution

Evolution means an orderly change. In biological science, the use of the word evolution usually refers to evolutionary theory. **Evolutionary theory** holds that modern-day plants and animals have developed from previous generations over a long period of time. All organisms living today have a common ancestor that evolved from the first living cells 3.5 billion years ago.

There is much evidence to support evolutionary theory. Part of that evidence is the appearance of homologous structures in different organisms. **Homologous structures** are body parts in different organisms that are similar even though they may function differently. For example, a human's arm, a whale's flipper, a bird's wing and a dog's foreleg are homologous structures. Even though at first glance they may appear different, the bones in each are actually so much alike that they have been given the same or similar names. According to evolutionary theory, the similarity of these structures implies that these organisms evolved from a common ancestor. The differences in structure are the result of adaptations to different environments.

Other evidence that supports evolutionary theory is the **embryonic similarity** between different organisms. An **embryo** is the early stage of development of an organism from a fertilized egg. The embryos of a fish, a bird and a human are similar. At one stage, all of these embryos even have gill slits and tail buds. As development continues, the differences become more dramatic. The early similarities of the embryos, however, imply that all three organisms inherited genes from a common ancestor.

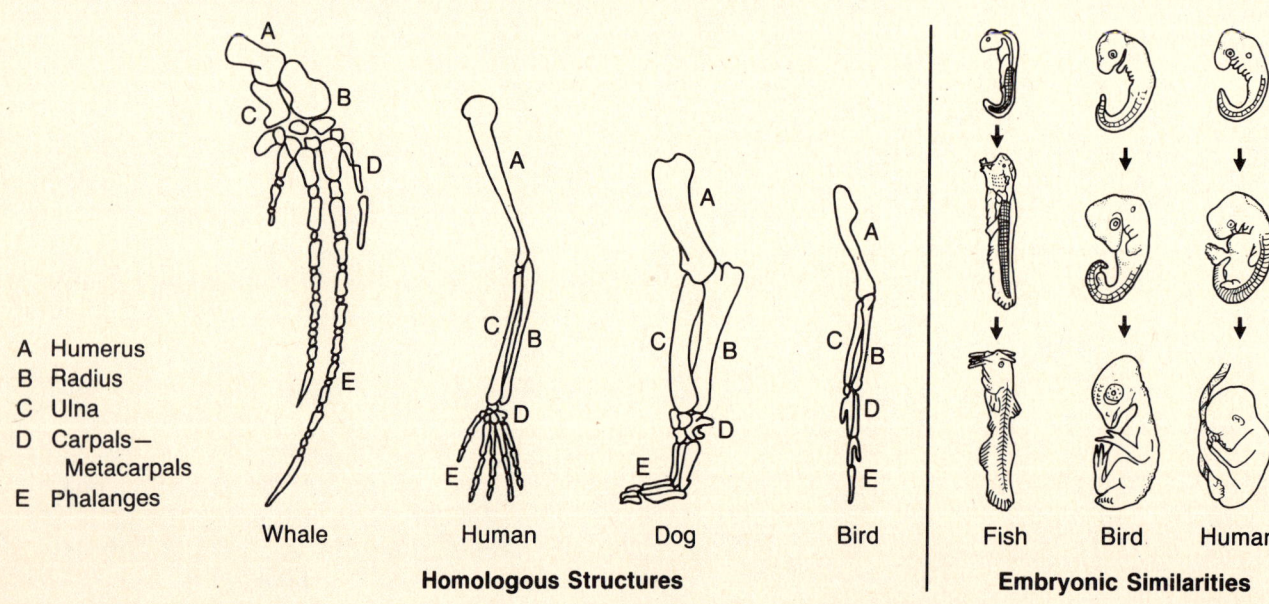

A Humerus
B Radius
C Ulna
D Carpals—Metacarpals
E Phalanges

Whale   Human   Dog   Bird

Homologous Structures

Fish   Bird   Human

Embryonic Similarities

Strategies for READING

Identify Cause and Effect Relationships

A cause and effect relationship indicates how one thing affects another. A cause is what makes something happen. An effect is what happens as a result.

Cause: What happened or what came first?
Effect: What was the result?

After reading the paragraph on page 28 the answers to these questions can be found.

Cause: Organisms developed homologous structures because they had a common ancestor.
Effect: Different organisms developed homologous structures.

The **cause and effect relationship** can also be stated as: The development of homologous structures (effect) is the result of a common ancestor (cause). The following key words or phrases are clues to cause and effect relationships: led to, the reason was, as a result of, have developed from, was caused by, due to, thus, therefore and so.

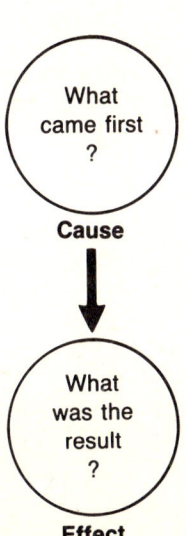

Examples

DIRECTIONS: Use the information on this and the preceding page to choose the one best answer for each item below.

1. What phrase is used to describe similar body parts in different organisms?

   (1) embryonic similarity
   (2) evolutionary theory
   (3) genetic pool
   (4) homologous structures
   (5) adaptation

Answer: (4) According to the article on page 28, homologous structures are body parts that are similar in different organisms. Each of the other items refers to something else. Choice (3) is not mentioned at all.

2. What do the early similarities of embryos imply about the origin of different organisms?

   (1) There is no common ancestor.
   (2) Genes were inherited from a common ancestor.
   (3) Evolutionary theory is disproven.
   (4) There is no genetic relationship.
   (5) Genes were not inherited.

Answer: (2) The article states that the early similarities of the embryos of different organisms (effect) implies that they inherited genes from a common ancestor (cause). Choices (1), (3), (4) and (5) contradict this idea.

Practice

HINT

As you are reading, look for cause and effect relationships. Identifying these relationships will help you better understand what you are reading. Refer to page 29 for some key words or phrases that identify these relationships.

DIRECTIONS: Choose the one best answer for each item below.

Items 1–4 refer to the following passage.

One of the key points in Darwin's theory of evolution is the theory of natural selection. According to this theory, those individuals within a species which have traits that help them adapt to the environment are more likely to survive and reproduce. The theory of **natural selection** is based on the following points:

1. Most organisms have more offspring than the environment is capable of supporting.
2. Because the number of individuals the environment can support is limited, they must compete for available resources.
3. The traits individuals within a species inherit vary a lot.
4. Those individuals who inherit traits that give them a competitive edge in their environment are better suited for survival.
5. Individuals that survive pass on traits to their offspring.

The result of natural selection is that organisms change as their environment changes over time. The evolution of an organism is, therefore, the result of natural selection.

1. Which is *not* part of Darwin's theory of evolution?

   (1) struggle for existence
   (2) natural selection
   (3) variations due to mutations
   (4) overproduction of organisms
   (5) competition for food and water

2. An organism with traits that help it adapt to its environment is *more* likely to

   (1) reproduce more offspring
   (2) change the environment
   (3) develop mutations
   (4) die from the environment
   (5) lose those traits

3. Which of the following is *most* likely the result of natural selection?

   (1) an insect's color
   (2) a person's language
   (3) an insect's name
   (4) an animal's home
   (5) an animal's food supply

4. Which of the following is necessary for natural selection to take place?

   (1) theory of evolution
   (2) length of lifespan
   (3) stability of environment
   (4) rate of growth
   (5) inheritance of traits

GO ON TO THE NEXT PAGE.

Items 5–8 refer to the following diagram and passage.

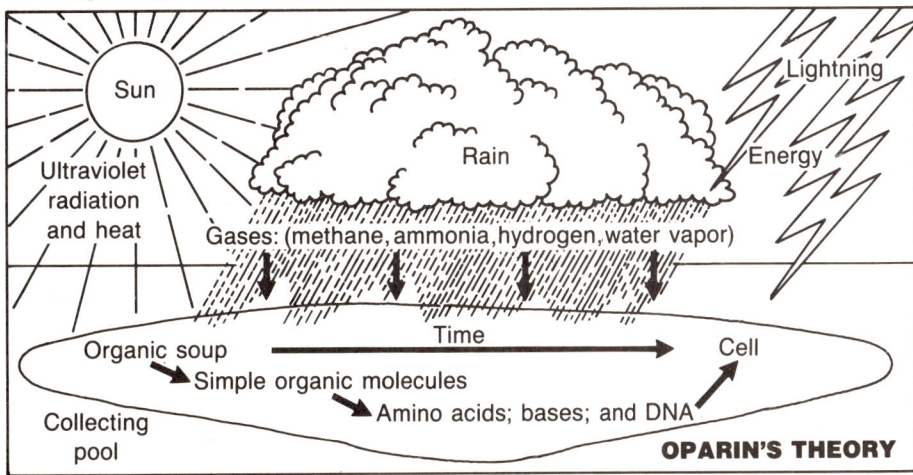

Life is thought to have first appeared about 3.5 billion years ago. Many scientists believe that at that time the earth's atmosphere was composed of the gases ammonia (NH_3), hydrogen (H_2), methane (CH_4) and water vapor (H_2O). These gases contain the elements carbon (C), nitrogen (N), hydrogen (H) and oxygen (O), the elements necessary to form organic molecules. **Organic molecules** all contain carbon in combination with one or more of these elements. Organic molecules are the building blocks of all living things.

According to one theory for the origin of life, energy from lightning or ultraviolet light from the sun split some of the gas molecules in the atmosphere. As a result, ions of carbon, nitrogen, hydrogen and oxygen were formed. Some of these ions recombined to form simple organic molecules. Some of these molecules were washed out of the atmosphere by rain and eventually collected in shallow pools on the surface of the earth. The organic molecules in these "organic pools" combined with one another to become more complex until the first living cells were formed.

5. A current theory concerning the origin of life assumes that the earth's primitive atmosphere contained

(1) ammonia, nitrogen and hydrogen
(2) methane, ammonia and water vapor
(3) carbon, methane and ammonia
(4) silicon, hydrogen and water
(5) nitrogen, hydrogen and water

6. According to the diagram, what was one of the complex organic molecules formed in the "organic pools"?

(1) hydrogen
(2) water
(3) carbon
(4) DNA
(5) ammonia

7. According to the article, which of the following caused some of the gas molecules to recombine in the early atmosphere?

(1) water vapor
(2) organic molecules
(3) methane ions
(4) amino acids
(5) lightning

8. It was necessary for organic molecules to exist before life could form because organic molecules

(1) make up all living things
(2) give rise to carbon
(3) provide energy
(4) make up the atmosphere
(5) produce rain

Before you take the GED Mini-Test, check your answers on pages 33–34.

GED Mini-Test 3

TIP When you are taking a test, some questions will seem more difficult than others. If you cannot answer a question right away, skip it and come back to the question later.

DIRECTIONS: Choose the one best answer for each item below.

Items 1–4 refer to the following passage.

The development of a new species from an old one is called speciation. The result of speciation is a new population of organisms that can no longer reproduce with their original ancestors. As long as the environment remains stable and the entire population of a species remains isolated, speciation is unlikely to occur. However, environments change and individual organisms may move from one area to another. In these situations, speciation is more likely to occur. One method by which speciation occurs is adaptive radiation.

Adaptive radiation occurs when a few individuals from a particular species became isolated from the general population. Isolation of this sort often occurs on islands or in areas isolated by mountains. These individuals must adapt to the pressures of the new environment and often develop new ways of life. The traits that help them adapt become dominant. Over a period of time, the isolated group gives rise to offspring that develop into new species in response to a particular part of the new environment. The results of adaptive radiation can be seen in Australia, where about 200 species of marsupials, such as the kangaroo and Koala bear, have all evolved from a common ancestor. The kangaroo adapted to life on the plains. The Koala bear adapted to life in the trees.

1. Which of the following is *most* likely to contribute to speciation?

   (1) environmental changes
   (2) environmental stability
   (3) isolation of an entire species
   (4) reproductive frequency
   (5) size of population

2. The marsupials of Australia are an example of

   (1) extinction
   (2) fossilization
   (3) convergent evolution
   (4) adaptive radiation
   (5) acquired traits

3. Which of the following words *best* describes the differences that occur in species during adaptive radiation?

   (1) convergent (similar)
   (2) random
   (3) divergent (changing)
   (4) individual
   (5) equal

4. How has speciation probably affected the diversity of animal and plant life?

   (1) created less diversity
   (2) not affected diversity
   (3) stopped diversity
   (4) slowed diversity
   (5) created more diversity

GO ON TO THE NEXT PAGE.

Items 5–6 refer to the following passage and diagram.

Genetic drift is the chance increase or decrease in the frequency of a particular gene occurring in a population. For example, suppose a population of squirrels consists mostly of brown squirrels with a few gray individuals. A waterway is constructed that divides the population into two isolated groups. The larger group still has several gray individuals, but the other, smaller group has only one gray individual that carries the gene for the gray trait. Before it can reproduce, the gray squirrel in the isolated group dies, leaving that population all brown. The gray gene has been lost to the isolated group's genetic pool. The frequency of the gray gene occurring has diminished to zero.

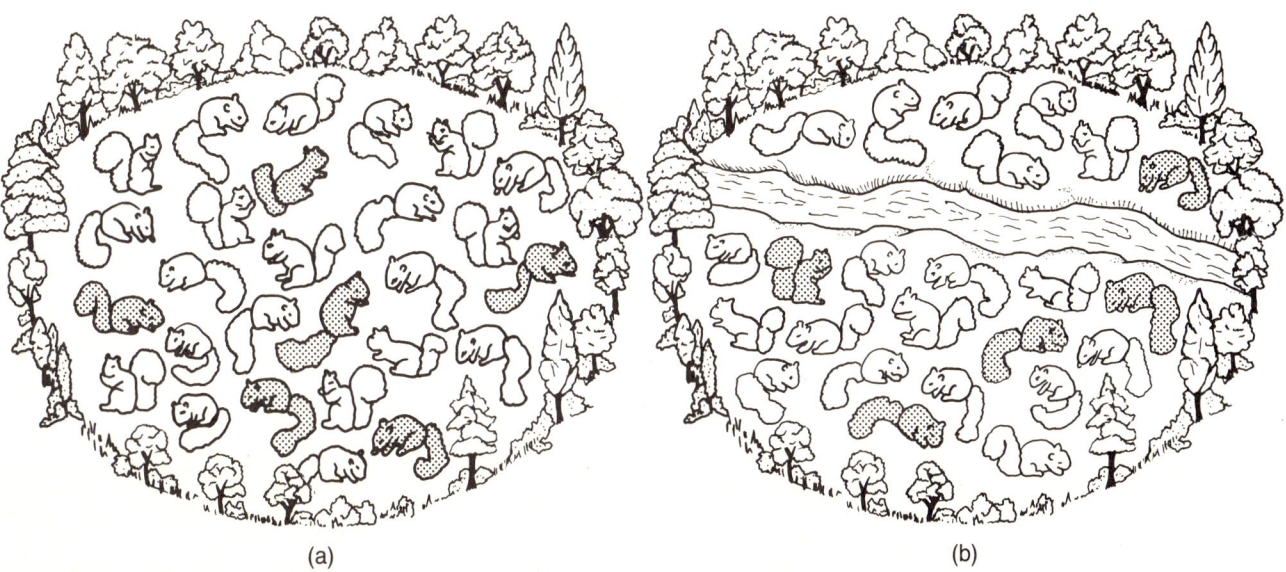

(a) (b)

5. Genetic drift is largely the result of

   (1) isolation of species
   (2) speciation
   (3) adaptive radiation
   (4) chance variation
   (5) gene pool

6. What sort of genetic drift can be expected among the larger population of squirrels due to the construction of the waterway?

   (1) increase the gray gene's frequency
   (2) decrease the gray gene's frequency
   (3) increase the brown gene's frequency
   (4) make the frequency of both genes equal
   (5) make the frequency of both genes constant

Check your answers to the GED Mini-Test on page 34.

Answers and Explanations

Practice   pp. 30–31

1. **Answer:** (3) Variations due to mutations is the only choice that is not mentioned in the article and did not develop as a theory until after Darwin's life-time. Choices (1), (2), (4) and (5) are all part of Darwin's theory; choices (1) and (5) are different ways of saying the same thing.

2. **Answer:** (1) Since an organism that adapts successfully to its environment is more likely to survive, it is more likely to reproduce. Choice (2) is not likely to be the result of adaptation. Choice (3) does not happen by adaptation, but rather by some mechanisms affecting the genetic make-up of an organism. Choices (4) and (5) state the opposite consequence of adaptation.

3. **Answer:** (1) The *most* likely result would be an insect's color, since color would probably have an effect on how well an insect adapts to its environment. Color is also an inherited trait that can be passed from one generation to another, a necessary condition of the natural selection process. Choices (2) through (5) are not the result of genetic inheritance and have nothing to do with an organism's adaptive ability.

5. **Answer:** (2) According to the article, the earth's atmosphere was composed of gases that together contained the elements carbon (C), nitrogen (N), hydrogen (H) and oxygen (O), the elements required to form organic compounds. Choices (1) and (3) do not contain oxygen (O), and choices (4) and (5) lack carbon (C).

7. **Answer:** (5) According to the article, energy from lightning was probably responsible for changing or splitting some of the gas molecules in the early atmosphere. Choice (1) washed organic molecules out of the atmosphere. Choice (2) was the result of ions recombining in the atmosphere. Choice (3) was the result of lightning splitting gas molecules in the atmosphere. Choice (4) was not present in the early atmosphere.

4. **Answer:** (5) Since natural selection favors those traits that help an organism to adapt to its environment, in order for the organism's offspring to survive, they must also inherit those traits. Natural selection is part of choice (1). Choices (2) and (4) may be affected by natural selection but are not necessary for it to take place. Choice (3) would diminish the process of natural selection.

6. **Answer:** (4) According to the diagram, DNA was probably formed in the "organic pools." Choices (1), (2) and (5) are not organic molecules. Choice (3) is an element found in all organic molecules. In addition, the diagram does not show these choices to be present in the "organic pools."

8. **Answer:** (1) According to the article, organic molecules are the building blocks of all living things. Therefore, organic molecules had to exist before living things could evolve. Organic molecules do not give rise to carbon, choice (2); carbon is part of any organic molecule. Energy, choice (3), was obtained from other sources before life emerged. Organic molecules are not part of the atmosphere, choice (4), gases are. Choice (5) is irrelevant.

GED Mini-Test pp. 32–33

1. **Answer:** (1) According to the article, environmental changes are one of the things that contribute to speciation. If choices (2) and (3) were true, speciation would be unlikely to occur. Choices (4) and (5) have little to do with speciation.

3. **Answer:** (3) During adaptive radiation, species diverge or change. They become dissimilar. Choice (1) is the opposite of divergent. Random differences, choice (2), are not a factor in adaptive radiation. Choice (4) does not *best* describe the differences that occur. Choice (5) is irrelevant.

5. **Answer:** (4) According to the article, genetic drift is the *chance* increase or decrease variation in the frequency of a particular gene occurring in a population. Choices (1) through (3) are used to describe ways in which new species may emerge. Choice (5) is affected by genetic drift.

2. **Answer:** (4) According to the article, the development of marsupials in Australia is a classic example of adaptive radiation. If choice (1) were true, there would be no marsupials. Choice (2) is irrelevant. Choice (3) is the opposite result of adaptive radiation. Choice (5) does not affect adaptive radiation.

4. **Answer:** (5) Since speciation allows new species of organisms to evolve, it creates more diversity. Choice (1) would be the result if no speciation occurred. Since choice (5) is correct, choices (2) and (3) cannot be correct. The rate of diversity, choice (4), is irrelevant to speciation.

6. **Answer:** (1) Since the larger population of squirrels now has more gray squirrels in proportion to reddish-brown squirrels, the frequency of the gray gene is greater. Choices (2) through (4) all reflect the wrong consequence of this ratio. Choice (5) is irrelevant.

Plant Growth

One of the characteristics of living things is their ability to grow. The growth in living organisms takes place internally, that is, by the internal division of cells. This sets them apart from inorganic, non-living substances, such as crystals, which grow by adding material externally from their environment.

The growth of seed plants takes place at specific places. These growing points are called **apical meristems** and are made up of rapidly dividing cells. The apical meristem at the tips of roots is covered by a **root cap,** which consists of a layer of thick, dead cells. The root cap protects the meristematic tissue as the root pushes its way through the soil.

Stems are divided into nodes and internodes. Meristematic tissue in the **node** gives rise to leaves, flowers and new branches. The area between nodes is called an **internode,** which displays little or no growth. Roots do not have nodes. Instead, meristematic tissue deep within the root may produce root branches anywhere along the root.

In woody plants, another kind of meristematic tissue called the cambium allows growth in the diameter or thickness of these plants. The **cambium** is a one-cell-thick, cylindrical layer in the stems, branches and roots. As the cells in the cambium layer divide, they add to the diameter of these structures.

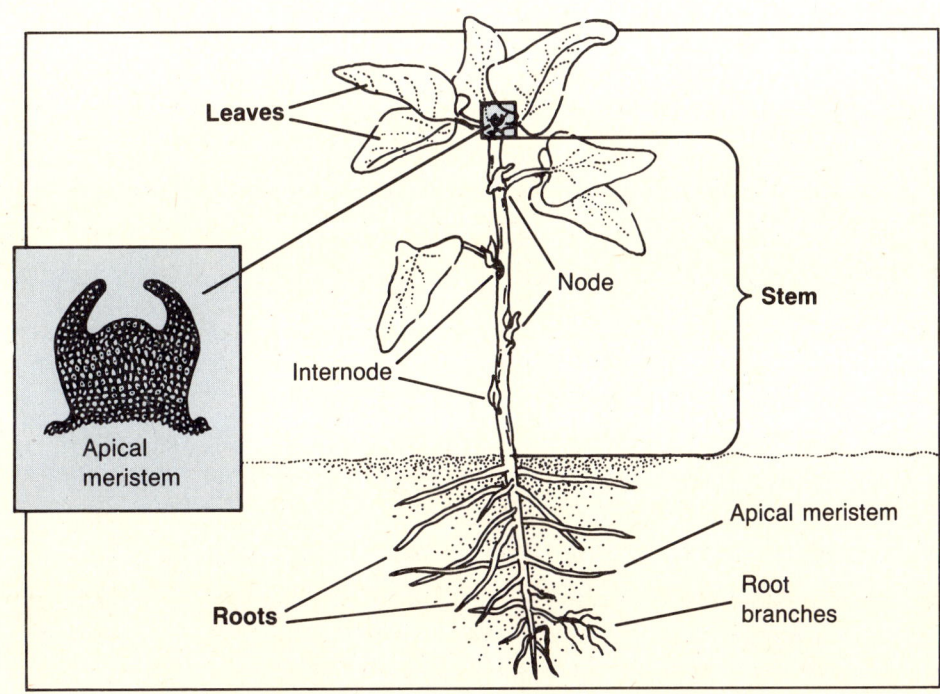

INTRODUCTION 35

Strategies for READING

Identify an Implication

This skill identifies assumptions, facts or statements that are taken for granted (not proved), and that the author takes for granted.

Sometimes an author omits certain facts or statements because he or she assumes the reader already knows them. The missing facts or statements are only **implied** by what the author has written. An author may also assume that certain conclusions are obvious from what has been written and do not need to be specifically stated. In such cases, the details in the paragraph or article imply the conclusion.

Read the second paragraph on page 35. Ask yourself, What kind of tissue makes up an apical meristem? The author never mentioned that an apical meristem is made up of meristematic tissue. Yet the fact is implied from the name "apical meristem" and from mention of meristematic tissue in the paragraph.

Now read the third paragraph. Ask, Do roots have internodes? Does the author answer this question, or is the answer implied? What detail implies that roots do not have internodes?

Examples

DIRECTIONS: Use the information on this and the preceding page to choose the <u>one</u> best answer for each item below.

1. What kind of meristematic tissue can be found in woody plants but *not* in most non-woody plants?

    (1) root cap
    (2) apical meristem
    (3) nodes
    (4) internodes
    (5) cambium

    Answer: (5)   According to the article, most woody plants have a cambium layer. Therefore, it is implied that most non-woody plants do not have cambium tissue. The article implies that choices (1) through (4) appear in non-woody plants as well as in woody plants.

2. If the meristematic tissue in a plant were destroyed, what *most* likely happen to the plant?

    (1) It would stop growing.
    (2) It would reproduce.
    (3) It would grow faster.
    (4) It would produce more leaves.
    (5) Its roots would start to branch.

    Answer: (1)   Since meristematic tissue is the growth tissue of a plant, if it were destroyed the plant would most likely stop growing (implied conclusion). Choices (2) through (5) all require growth from meristematic tissue.

READING: COMPREHENSION

Practice

Look for implied facts, statements and conclusions. Recognizing implications will help you understand what you are reading.

DIRECTIONS: Choose the one best answer for each item below.

Items 1–4 refer to the following paragraph and chart.

The study of the classification of organisms is called **taxonomy**. Taxonomy is based on the evolutionary relationships between different groups of organisms.

Kingdom	Characteristics	Examples
Monera	Single-celled; nucleus and organelles lack membranes; nutrition by absorption; reproduction by fission	Viruses, bacteria, blue-green algae
Protista	Both single-celled and multicellular; organized nucleus and organelles with membranes	*Ameba, Paramecium,* slime molds, algae
Fungi	Firm cell walls; both single-celled and multicellular forms; reproduction by budding, fragmentation or sexual; most have a body formed by the massing of filaments	Mushrooms, bread molds, yeasts, smuts, bracket fungi
Plantae	Multicellular organisms with tissues and organs; cell walls contain cellulose; cells contain chlorophyll; sex organs are multicellular; autotrophic	Liverworts, ferns, mosses, club mosses, flowering plants, trees
Animalia	Multicellular organisms with complex tissues and organ systems; most reproduce sexually; no chlorophyll or cellulose; eukaryotic, heterotrophic organisms	Insects, fish, amphibians, reptiles, birds, mammals

1. What is the relationship between taxonomy and evolution?

    (1) Evolution is based on taxonomy.
    (2) Taxonomy is based on evolution.
    (3) Evolution and taxonomy are the same thing.
    (4) Evolution is the study of taxonomy.
    (5) Taxonomy is the study of evolution.

2. Bread mold, which is formed by the compilation of long filaments, is an example of which kingdom?

    (1) Monera
    (2) Protista
    (3) Fungi
    (4) Plantae
    (5) Animalia

3. Which characteristic is typical of *only* Monera?

    (1) no nuclear membrane
    (2) reproduce by budding
    (3) single-celled
    (4) organelles with membranes
    (5) sexual reproduction

4. A fern, with its rigid cell walls, would have which of the following characteristics?

    (1) organelles lacking membranes
    (2) absence of chlorophyll
    (3) cell membrane but no cell wall
    (4) nucleus lacking a membrane
    (5) contain chlorophyll

GO ON TO THE NEXT PAGE.

Items 5–6 refer to the chart on page 37.

5. *Trypanosoma* is a single-celled organism with a nuclear membrane and no cell wall. To which kingdom would *Trypanosoma* belong?

   (1) Monera
   (2) Protista
   (3) Fungi
   (4) Plantae
   (5) Animalia

6. A bacterium, which absorbs nutrition through its single cell wall, would be classified as a

   (1) Monera
   (2) Protista
   (3) Fungi
   (4) Plantae
   (5) Animalia

Items 7–10 refer to the following chart.

Class	Characteristics of Vertebrates	Examples
Agnatha	Eel-like freshwater, marine animals without true jaws, scales, fins or skeletons	lamprey, hagfish
Chondrichthyes	Fishes with true jaws and fins; gills open through gill slits; cartilaginous skeleton	Sharks, rays, skates
Osteichthyes	Freshwater, marine fishes; bony skeleton; free-moving gills; true jaws and fins	All bony fishes
Amphibia	Slimy skin; limbs without claws; metamorphosis; three-chambered heart; external fertilization	Salamanders, frogs, newts, toads
Reptilia	Scale-covered body; toes with claws; internal fertilization	Turtles, tortoises, lizards, snakes
Aves	Body covered with feathers; wings; four-chambered heart; warm-blooded	All birds
Mammalia	Body covered with hair; four-chambered heart; mammary glands; highly developed cerebrum and cerebellum	Kangaroo, horse, bat, cat, dog, all primates

7. If you found an organism in the ocean that had jaws, fins and a bony skeleton, which class would you put it in?

   (1) Agnatha
   (2) Osteichthyes
   (3) Chondrichthyes
   (4) Reptilia
   (5) Aves

8. To which class does a bald eagle belong?

   (1) Chondrichthyes
   (2) Amphibia
   (3) Reptilia
   (4) Aves
   (5) Mammalia

9. Which of the following characteristics is typical of *only* mammals?

   (1) body covered with hair
   (2) four-chambered heart
   (3) toes with claws
   (4) warm-blooded
   (5) bony skeleton

10. Which of the following characteristics would you expect a catfish to have?

    (1) four-chambered heart
    (2) bony skeleton
    (3) mammary glands
    (4) internal fertilization
    (5) highly developed cerebrum

Before you take the GED Mini-Test, check your answers on pages 40–41.

GED Mini-Test

TIP 4

A key to passing the GED test is to read for understanding. Reading the test questions before reading a passage can help you know what you will be reading. They often help provide a summary.

DIRECTIONS: Choose the one best answer for each item below.

Items 1–6 refer to the following passage.

All seed plants are divided into two groups: gymnosperms and angiosperms. Gymnosperms are non-flowering seed plants. The name gymnosperm means "naked seed," which refers to the fact that the seeds develop exposed without any protective covering around them. There are about 750 species living today. The most common gymnosperms include the pines, spruces, hemlocks, junipers and yews.

Angiosperms, on the other hand, include all flowering plants. The biological success of angiosperms has been remarkable. They can be found in almost every kind of climate—from hot, dry deserts to cold mountain tops. In all, there are about 235,000 recognized species of angiosperms. The success of the angiosperms can be traced to several factors. Genetic variations have produced many different forms of trees, shrubs, herbs and plants that can grow in a variety of environments and thrive in different soils and temperatures. Besides the genetic variations, the seeds, flowers and fruits of angiosperms have also contributed to their success. Angiosperms depend on wind, insects or water to pollinate their flowers. Their seeds are enclosed in a protective covering such as a pod or the fleshy part of a fruit. And the seeds contain stored food that nourishes the young plant until it is established on its own.

All angiosperms are divided into two classes: Monocotyledons and Dicotyledons. The main basis of the division is the number of cotyledons, or seed leaves, that develop as part of the embryo plant. A cotyledon is the first leaf or leaves of the embryo plant. It contains food that nourishes the young seedling until photosynthesis begins. A monocot has only one cotyledon, while a dicot has two cotyledons. Besides the differences in the number of seed leaves, monocots and dicots have two other differences: the leaves of monocots have parallel veins and the flower parts are usually in sets of three, whereas the leaves of dicots have net-like veins and the flower parts are usually in sets of four or five. Most woody trees, shrubs and herbs are dicots. Examples of monocots include palms and most non-woody plants, such as grasses, lilies, irises and orchids. Farm crops include examples from both classes. Corn and most grain plants are monocots. Tomatoes, potatoes and most beans are dicots.

GO ON TO THE NEXT PAGE.

1. Why have spruce trees not been nearly as successful as palms? Spruce trees

    (1) have flowers
    (2) have exposed seeds
    (3) have protected seeds
    (4) have two cotyledons
    (5) have one cotyledon

2. Which of the following would *not* have net-like veins?

    (1) potatoes
    (2) lima beans
    (3) string beans
    (4) tomatoes
    (5) wheat

3. Monocots differ from dicots in that monocots have

    (1) needle-shaped leaves
    (2) leaves with net-like veins
    (3) a single cotyledon
    (4) two cotyledons
    (5) flower parts in sets of four

4. If you found a flower with five petals, to which class of flowering plants would it probably belong?

    (1) gymnosperms
    (2) angiosperms
    (3) monocotyledons
    (4) dicotyledons
    (5) anthophyta

5. Since wheat belongs to the family of grasses, you would expect a wheat plant to have

    (1) flower parts in sets of four or five
    (2) flower parts in sets of three
    (3) net-like veins
    (4) two cotyledons
    (5) unprotected seeds

6. Which of the following probably *best* explains the remarkable biological success of angiosperms?

    (1) They produce unprotected seeds.
    (2) They have net-like veins.
    (3) They produce protected seeds.
    (4) They produce reproductive cones.
    (5) They produce seeds with one cotyledon.

Check your answers to the GED Mini-Test on page 41.

Answers and Explanations

Practice   pp. 37–38

1. **Answer:** (2)   According to the paragraph, taxonomy is based on the evolutionary relationships between groups of organisms. Choice (1) states the opposite. Choices (3) through (5) are incorrect since taxonomy is the study of the classification of animals and evolution is the study of the development of organisms.

2. **Answer:** (3)   Since bread mold is formed by the compilation of long filaments, it must be an example of fungi; only fungi have such a characteristic. Choices (1), (2), (4) and (5) are kingdoms that do not have organisms with filamentous bodies.

3. **Answer:** (1)   According to the chart, only those organisms belonging to the kingdom Monera do not have nuclear membranes. Choice (2) is characteristic of Monera and Fungi. Choice (3) is also characteristic of Protista. Choice (4) is not present in Monera, but is present in the organisms that belong to the other kingdoms. Choice (5) is a characteristic of Fungi, Plantae and Animalia.

4. **Answer:** (5)   Since ferns have cells with rigid cell walls, they must belong to the kingdom Plantae and organisms in Plantae have cells containing chlorophyll. All other items refer to cellular characteristics typical of organisms found in other kingdoms.

5. **Answer:** (2) According to the chart, Protista are single-celled organisms that have nuclear membranes. Cell walls are characteristic of Fungi and Plantae, choices (3) and (4). All other choices are names of other kingdoms.

6. **Answer:** (1) According to the chart, Monera are organisms that get their nutrition mostly by absorption. Therefore, a bacterium must be classified in the kingdom Monera. None of the other choices (2) through (5) have that characteristic.

7. **Answer:** (2) According to the chart, organisms belonging to the class Osteichthyes are characterized by having jaws and fins and a bony skeleton. Choice (3) has jaws and fins but a cartilaginous skeleton. All other choices name classes of organisms that have different characteristics.

8. **Answer:** (4) Since bald eagles are birds, they must have the characteristics of a "body covered with feathers; wings . . . " etc., and, therefore, belong to the class Aves. All other choices name classes that include other types of animals.

9. **Answer:** (1) According to the chart, one of the characteristics of mammals that is not shared with any other class of animals is a body covered with hair. Choices (2) through (5) name characteristics that are not specific only to mammals.

10. **Answer:** (2) A catfish is obviously a fish, and only choice (2) is a characteristic of bony fishes. Choices (1), (3) and (5) are characteristics of the class Mammalia. Choice (4) is mentioned as a characteristic of Reptilia, and is also found in Aves and Mammalia.

GED Mini-Test pp. 39–40

1. **Answer:** (2) According to the article, spruce trees are an example of a gymnosperm, and gymnosperms have "exposed seeds," which makes their survival more difficult. The palm is an angiosperm, and it has protected seeds, choice (3), and flowers, choice (1). Choices (4) and (5) represent differences between two kinds of angiosperms.

2. **Answer:** (5) According to the article, net-like veins are characteristics of dicot leaves, but wheat is a grass plant and the article states that grasses are monocots. So wheat would *not* have net-like veins, but would have the parallel veins of a monocot. Choices (1) through (4) are examples of dicots and therefore *do* have net-like veins.

3. **Answer:** (3) According to the article, one way in which monocots differ from dicots is that monocots have a single cotyledon. Choice (1) names a characteristic of conifers. Choices (2), (4) and (5) name characteristics of dicots.

4. **Answer:** (4) According to the article, one of the characteristics of dicotyledons is that their flower parts are usually in sets of four or five. Choices (1) and (2) name the two groups into which all seed plants are divided. Monocotyledons, choice (3), have flower parts in sets of three. Choice (5) names the phylum to which all angiosperms belong.

5. **Answer:** (2) Since wheat belongs to the grass family, and grasses are monocots, you would expect the parts of a wheat plant flower to be arranged in sets of three. Choices (1), (3) and (4) name characteristics of dicots. Choice (5) names a characteristic of gymnosperms.

6. **Answer:** (3) Since the seeds of angiosperms are protected by some kind of covering, it can be assumed that has contributed to their remarkable biological success. Choices (1) and (4) name characteristics of gymnosperms. Choice (2) is characteristic of only dicots, which have been quite successful. Choice (5) is characteristic of only monocots, which have also been very successful.

ANSWERS AND EXPLANATIONS

5 The Human Digestive System

The digestive system consists of one continuous tube called the **alimentary canal,** made up of the mouth, esophagus, stomach, small intestine and large intestine as well as the **liver** and the **pancreas.** They are connected to the alimentary canal by tubes called **ducts.**

Mouth and esophagus The mouth prepares food for digestion. In the mouth food is ground up and mixed with saliva. Saliva is about 95% water, but it also contains the enzyme **amylase** that begins to break down the starches in food. Mucus in saliva lubricates the food so that it can pass down the esophagus easily. Involuntary contractions of the walls of the esophagus keep the swallowed food moving in the right direction.

Stomach The lining of the stomach contains three different kinds of glands. One kind secretes the enzyme **pepsin,** which begins to break down protein. Another kind secretes **hydrochloric acid,** needed to activate the pepsin enzyme and also dissolve minerals and kill bacteria that enter the stomach with food. The third kind of gland secretes **mucus,** which protects the stomach lining from the acid in the stomach.

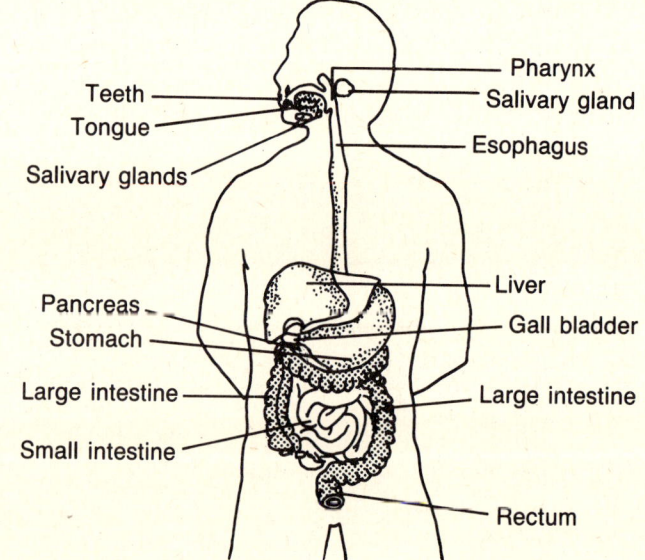

Small intestine The small intestine is about three centimeters in diameter and seven meters in length and is where most of the digestive process takes place. The lining of the small intestine contains many glands that secrete enzymes that break down complex sugars and further break down proteins.

The **pancreas** secretes into the stomach enzymes that break down proteins, starches and fats. The **liver** secretes bile, which is stored in the gall bladder until it is needed by the stomach. Bile breaks up large fat globules into smaller droplets.

Once the digestive process has been completed in the small intestine, nutrients from the food diffuse, or pass through, the lining of the small intestine and into the blood stream. The remaining undigested material passes into the large intestine.

Large intestine The undigested material from the small intestine is mixed with a large amount of water. One of the main functions of the large intestine is to absorb this water. As the water is absorbed, the undigested material becomes more solid and eventually passes out of the large intestine through the anal opening.

Strategies for READING

Recognize Unstated Assumptions

This skill involves identifying facts or statements that an author takes for granted (does not prove).

In order to understand and recognize assumptions an author may make, you must carefully read all the information presented. It is similar to inferring the unstated main idea of a paragraph. Identify the topic, carefully follow the author's logic and decide what point the supporting details make about the topic.

Read the second paragraph of the article on page 42. The author states that in the mouth, food is ground up and mixed with saliva. It is never stated how the food is ground up. The author assumes the reader knows food is ground up by the teeth.

After reading the entire article on page 42, you may recognize another assumption the author has made. Ask yourself, What is the beginning and the end of the alimentary canal? The author never answers this question. Yet it can be assumed from the information in the article that the beginning of the alimentary canal is the mouth and the end is the anal opening.

Read the third paragraph of the article. Where does acid in the stomach come from? The author assumes you will know after reading the paragraph. What information in the paragraph tells you?

Examples

DIRECTIONS: Use the information on this and the preceding page to choose the <u>one</u> the best answer for each item below.

1. In which direction do the esophageal contractions move swallowed food?

    (1) toward the mouth
    (2) toward the stomach
    (3) to one side
    (4) past the gall bladder
    (5) in front of the pancreas

Answer: (2) Food passes from the mouth through the esophagus to the stomach (unstated assumption). If choice (1) were correct we could never swallow. The esophagus does not go to one side, choice (3). Choices (4) and (5) are not mentioned in the article.

2. Nutrients from the small intestine enter the blood stream and are carried to the

    (1) large intestine
    (2) pancreas
    (3) liver
    (4) gall bladder
    (5) rest of the body

Answer: (5) The blood stream carries the nutrients throughout the body (unstated assumption). Though blood carries nutrients to choices (1) through (4), each of those items is incomplete by itself. Choice (5) includes choices (1) through (4) and the rest of the body.

Practice

Look for assumptions while you are reading. Identifying assumptions will help you better understand the points an author is trying to make.

DIRECTIONS: Choose the <u>one</u> best answer for each item below.

Items 1–6 refer to the following passage and diagram.

The **heart** is the major organ of the circulatory system. It is located under the breast bone and behind the lungs and is about the size of a fist. The heart beats on the average of 72 times a minute, pumping blood throughout the body.

The right and left sides of the heart are divided by a wall called the **septum.** Each side is divided into two chambers. The upper chambers are called the **atria,** and the lower chambers are called the **ventricles.**

Oxygen-depleted blood from the body enters the right atrium through a large vein called the **vena cava.** When the atrium is full, it contracts, forcing the blood through a valve into the right ventricle. The valve closes to keep the blood from flowing back into the atrium. Next, the right ventricle contracts, forcing the blood into the pulmonary arteries, which carry it to the lungs. In the lungs the blood picks up oxygen and flows back to the heart where it enters the left atrium through **pulmonary veins.** When the left atrium is full, it contracts and forces the blood through a valve into the left ventricle. As the left ventricle contracts, it forces the blood into a large artery called the **aorta.** From the aorta, blood flows into a system of blood vessels that carry it throughout the body.

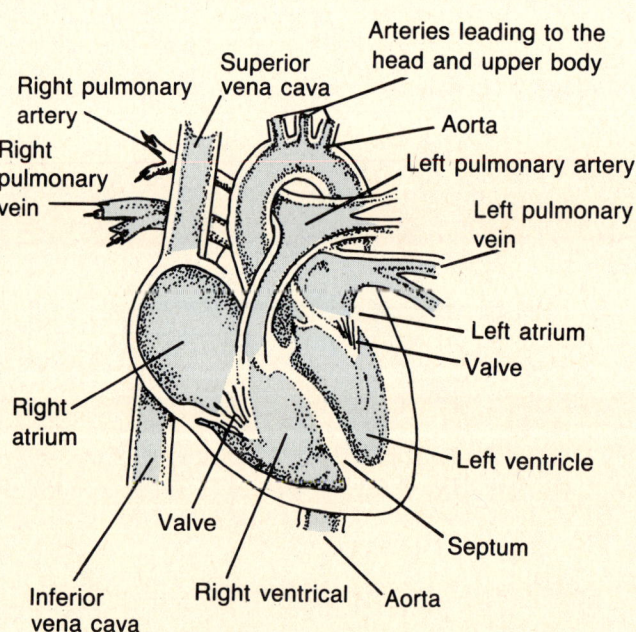

1. About how many times does the average heart beat each hour?

   (1) 72
   (2) 504
   (3) 1,728
   (4) 4,320
   (5) 10,680

2. Which heart chambers contain blood with the *highest* concentration of oxygen?

   (1) left atrium, right atrium
   (2) left atrium, left ventricle
   (3) left ventricle, right ventricle
   (4) right atrium, right ventricle
   (5) right atrium, left ventricle

GO ON TO THE NEXT PAGE.

3. Which chamber is responsible for sending blood from the heart through the body?

   (1) right ventricle
   (2) right atrium
   (3) left ventricle
   (4) left atrium
   (5) vena cava

4. What would probably happen if the valves between the atria and the ventricles did not completely close?

   (1) Blood would not enter atria.
   (2) Blood would not enter vena cava.
   (3) Blood would flow back into atria.
   (4) Heart would not contract.
   (5) Aorta would receive too much blood.

5. What would probably happen if the vena cava became clogged?

   (1) Too much blood would enter the right atrium at once.
   (2) Too little blood would enter the right atrium at once.
   (3) Too much blood would enter the left atrium at once.
   (4) Too much blood would enter the lungs at once.
   (5) Too much blood would enter the aorta at once.

6. What is *most* likely to happen if the aorta were pinched off?

   (1) Too much blood would enter the left ventricle.
   (2) Not enough oxygen-depleted blood would enter the heart.
   (3) Oxygen-depleted blood would not be able to reach the lungs.
   (4) The body would not get enough oxygenated blood.
   (5) Oxygenated blood would enter the vena cava.

Items 7–10 refer to the following passage.

Arteries carry blood away from the heart. They divide into smaller vessels called **arterioles,** which enter the body's tissues and branch into **capillaries.** The walls of capillaries are only one cell thick. In this way dissolved nutrients diffuse, or pass through, the thin capillary walls into the body's cells and waste products diffuse out of the tissues and into the capillaries to be carried away. Capillaries form **venules,** which combine to form the large blood vessels called **veins.** Veins carry the blood back to the heart.

7. In which of the following blood vessels would you expect waste products to be carried?

   (1) veins
   (2) arterioles
   (3) arteries
   (4) aorta
   (5) heart

8. Which blood vessel carries blood from the heart to the lungs?

   (1) vein
   (2) venule
   (3) artery
   (4) capillary
   (5) lymph

9. Which blood vessel carries blood from the kidneys to the heart?

   (1) vein
   (2) venule
   (3) artery
   (4) capillary
   (5) lymph

10. Blood in the pulmonary veins would be rich in

    (1) carbon dioxide only
    (2) oxygen only
    (3) oxygen and carbon dioxide
    (4) neither oxygen nor carbon dioxide
    (5) waste material

Before you take the GED Mini-Test, check your answers on pages 47–48.

GED Mini-Test 5

TIP Answer *all* test questions. Remember that you are not penalized for a wrong answer on the GED test. So if you do not know the answer, *guess*.

DIRECTIONS: Choose the one best answer for each item below.

Items 1–6 refer to the following passage, chart and diagram.

The endocrine system is made up of various endocrine glands and the hormones they secrete. Hormones are substances that affect specific cells in different parts of the body. Unlike exocrine glands, which secrete substances through ducts, endocrine glands release their hormones directly into the blood stream. Review the diagram and the chart to learn the location of the endocrine glands and the effects that some of their hormones have on the body.

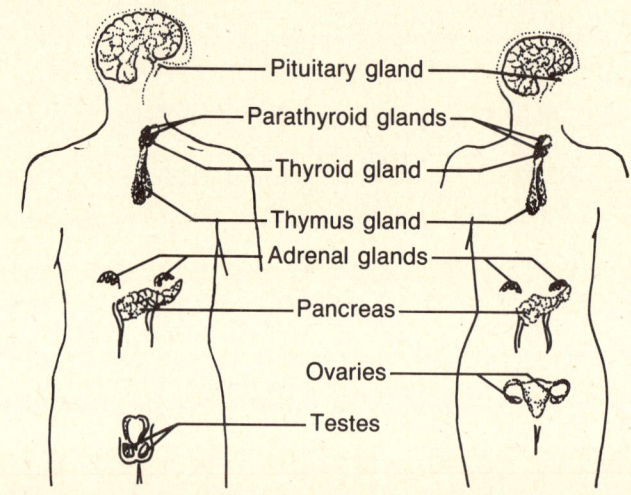

Endocrine Gland	Hormone	Effect
Thyroid	Thyroxin	Controls how fast food is converted to energy in cells
Parathyroid	Parathormone	Regulates body's use of calcium and phosphorus
Thymus	Thymosin	May affect the formation of antibodies in children
Adrenal	Adrenaline	Prepares the body to meet emergencies
	Cortisone	Maintains salt balance
Pancreas	Insulin	Decreases level of sugar in the blood
Ovaries (female gonads)	Estrogen	Controls the development of secondary sex characteristics
Testes (male gonads)	Testosterone	Controls the development of secondary sex characteristics
Pituitary	Growth hormone	Controls the growth of bones and muscles
	Oxytocin	Causes uterine contractions in labor
	ACTH, TSH, FSH, LH, LTH	Regulate the secretions of the other endocrine glands

GO ON TO THE NEXT PAGE.

1. The body normally responds to high concentrations of sugar in the blood by secreting

   (1) glucagon
   (2) insulin
   (3) estrogen
   (4) testosterone
   (5) adrenaline

2. Which endocrine gland is located at the base of the brain?

   (1) thymus
   (2) pancreas
   (3) thyroid
   (4) parathyroid
   (5) pituitary

3. Which endocrine gland is different in males and females?

   (1) pituitary
   (2) gonads
   (3) adrenal
   (4) thymus
   (5) pancreas

4. A person with high levels of sugar in the blood is said to have diabetes. Which hormone is lacking in a person with this condition?

   (1) thyroxin
   (2) adrenaline
   (3) insulin
   (4) oxytocin
   (5) thymosin

5. A condition called acromegaly causes certain bones in the adult to grow larger and thicker. Which endocrine gland is responsible for this condition?

   (1) adrenal
   (2) pancreas
   (3) thymus
   (4) thyroid
   (5) pituitary

6. When you are confronted with an emergency, your heart rate and breathing quicken and you have a sudden burst of energy. Which endocrine gland is responsible for this reaction?

   (1) adrenal
   (2) thyroid
   (3) thymus
   (4) parathyroid
   (5) pancreas

Check your answers to the GED Mini-Test on page 48.

Answers and Explanations

Practice pp. 44–45

1. **Answer:** (4) According to the passage, the average heart beats 72 times a minute. To determine how many times it would beat in one hour, multiply 72 beats per minute by 60 minutes in a hour. Choice (1) is the number of times the average heart beats in one minute. Choices (2) and (3) are irrelevant. Choice (5) is the number of times the average heart beats in 24 hours.

2. **Answer:** (2) Oxygen-rich blood enters the heart from the pulmonary veins into the left atrium and is then pumped into the left ventricle. Blood in the right atrium and right ventricle, choices (1), (3), (4) and (5), is oxygen-depleted since it has entered the heart from the vena cava, which carries carbon dioxide and waste from all parts of the body back to the heart.

3. **Answer:** (3) The left ventricle forces blood into the aorta, which directs it throughout the body. Choice (1) sends deoxygenated blood to the lungs. Choice (2) collects deoxygenated blood from the body. Choice (4) receives oxygenated blood from the lungs. Choice (5) is not one of the chambers of the heart, but a blood vessel.

4. **Answer:** (3) The valves between the atria and ventricles prevent the blood in the ventricles from flowing back into the atria. Choices (1), (2) and (4) would still occur. The aorta, choice (5), would if anything receive less blood since less blood would flow through the heart because of a faulty valve.

5. **Answer:** (2) Since the vena cava brings blood from the body to the right atrium, a clogged vena cava would slow the rate of blood entering the right atrium. Choice (1) states the opposite. The vena cava does not connect with choices (3) and (4). The slowed flow of blood to the right atrium would also slow the flow of blood to the aorta, choice (5).

6. **Answer:** (4) Since the aorta sends oxygenated blood on its way throughout the body, if it were pinched off this flow of blood would be restricted. Choice (1) is not connected to the aorta. Oxygen-depleted blood, choice (2), enters the heart through the vena cava not the aorta. Oxygen-depleted blood, choice (3), passes to the lungs through the pulmonary arteries not the aorta. Oxygenated blood does not enter the vena cava, choice (5).

7. **Answer:** (1) According to the paragraph, waste products are carried away from the body's tissues and veins carry blood from the tissues back to the heart. Choices (2) through (4) all refer to arteries. Choice (5) is not a blood vessel.

8. **Answer:** (3) According to the paragraph, arteries carry blood away from the heart. Choices (1) and (2) refer to veins. Choice (4) connects arteries and veins. Choice (5) refers to a different part of the circulatory system and is irrelevant.

9. **Answer:** (1) The kidneys produce waste products that diffuse out of the body tissues into capillaries, venules and veins to be carried back to the heart. Choices (2) and (4) are found in the body tissues. Choice (5) is not mentioned, and choice (3) carries blood away from the heart.

10. **Answer:** (2) Only the pulmonary veins coming from the lungs are rich in oxygen. Choices (1) and (3) are incorrect because carbon dioxide is given off in the lung capillaries. Choice (4) is absurd, and choice (5) is carried by arteries to the kidney for excretion.

GED Mini-Test *pp. 46–47*

1. **Answer:** (2) According to the chart, insulin is the hormone that decreases the blood-sugar level. Choice (1) is not mentioned in the chart. Choices (3) and (4) are the male and female sex hormones. Choice (5) is the hormone secreted in emergencies.

2. **Answer:** (5) According to the diagram, the only endocrine gland associated with the brain is the pituitary gland. Choices (1) through (4) are located below the neck and cannot, therefore, be located at the base of the brain.

3. **Answer:** (2) According to the chart, the gonads are different in males and females. In males they are called testes and secrete mainly testosterone. In females they are called ovaries and secrete mainly estrogen. Choices (1) and (3) through (5) name other endocrine glands that are common to both males and females.

4. **Answer:** (3) According to the chart, insulin secreted by the adrenal glands lowers the level of sugar in the blood. Therefore, if the blood has a high level of sugar, insulin must be lacking in adequate amounts. Choices (1), (2), (4) and (5) name other hormones that perform different functions.

5. **Answer:** (5) According to the chart, growth hormone produced by the pituitary gland controls the growth and development of bones and muscles. A sudden growth in some bones, therefore, is probably the effect of growth hormone produced by the pituitary gland. Choices (1) through (4) name other endocrine glands serving different functions from the one described in the question.

6. **Answer:** (1) According to the chart, the adrenal glands prepare the body for emergencies. The reactions stated in the question are reactions to emergency situations and are controlled at such times by the adrenal glands. Choices (2) through (5) name other endocrine glands.

6 Defenses Against Disease

Diseases that can be spread from one person to another are called **infectious diseases.** Infectious diseases are caused by microorganisms called **pathogens.** The most common kinds of pathogens are **bacteria** and **viruses.**

The body has several defenses against infectious diseases. One line of defense is mostly mechanical. It prevents disease organisms from entering the tissues of the body. Another line of defense is mostly biological. It reacts to disease organisms that have managed to get inside the body's tissues.

Mechanical defenses The skin is the first barrier to disease organisms. The outer layer of the skin, called the **epidermis,** is made up of dead cells that stop bacteria from entering the tissues that lie beneath the skin. If there is a break in the skin from a cut or some other wound, a blood clot forms to close the wound. Sometimes, however, a wound is too large for a blood clot to close effectively. When this happens, bacteria may enter the body through the wound.

Mucus membranes line many parts of the body, such as the nasal passages, the lungs and the digestive tract. **Mucus** is a thick, clear, slippery fluid secreted by mucus cells. **Mucus membranes** help to protect the body's tissues from invasion by disease organisms. Mucus linings in the digestive tract help keep disease organisms from entering the body through the walls of the stomach and intestines. Mucus linings in the nasal passages and passages of the lungs trap dust particles and bacteria, which are then passed out through the throat and mouth.

Biological defenses When disease-causing microorganisms enter the body, white blood cells called **phagocytes** seek them out and engulf them (see illustration at right). These cells pass through capillary walls and into the body's tissues. One kind of phagocyte can enlarge to become a **macrophage.** These cells are so large they can engulf a hundred or more bacteria at one time.

Other kinds of white blood cells called **lymphocytes** produce antibodies. **Antibodies** are protein substances that react with specific foreign materials in the body and make them ineffective. Lymphocytes can produce many different antibodies to attack different kinds of disease organisms.

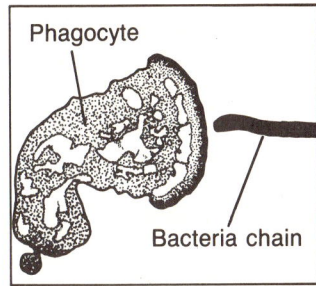
↑0

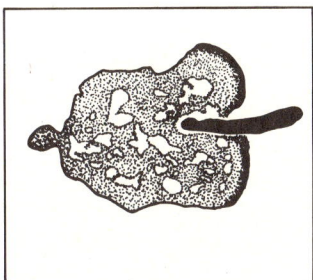
30 sec.

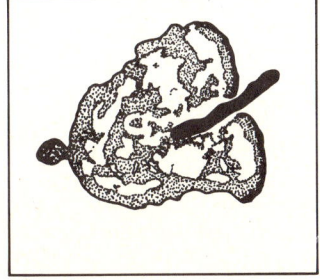
50 sec.

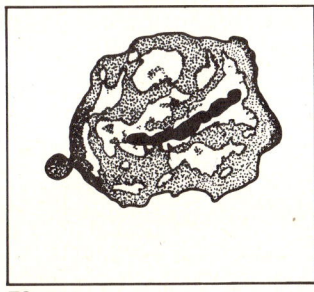
70 sec.

Phagocyte Engulfing Chain of Bacteria

INTRODUCTION 49

Strategies for READING

Recognize the Role of Values and Beliefs

This skill involves identifying an author's prejudice, making a judgment about what is written and determining the accuracy of available data to substantiate conclusions.

The way an author states things may be influenced by his or her prejudices. However, your own feelings should not influence your choice of answers. For example, if a test-taker is a strict vegetarian but the correct answer to a passage on nutrition is—Adults should eat four helpings of meat a week—then that is the *correct* answer, even if your tendency is to eat salads.

Read the first paragraph of the passage on page 49. It states that infectious diseases are caused by microorganisms. You may believe psychological and emotional factors cause infectious disease. But read the rest of the passage. Are the facts presented and does the data substantiate that conclusion? Remember the author's point of view must be taken into consideration when assessing the information. Words that will help you identify a belief are believe, feel, seem and could be.

Examples

DIRECTIONS: Use the information on this and the preceding page to choose the one best answer for each item below.

1. Cigarette smoke damages the mucus lining of the lungs. This could result in an increased risk of

   (1) antibody damage
   (2) becoming addicted to smoking
   (3) blood clotting
   (4) phagocyte damage
   (5) contracting infectious diseases

   Answer: (5) According to the article, the mucus lining of the lungs protects the body from infectious diseases. If this lining is damaged, it stands to reason that an increased risk of contracting infectious diseases might be the result.

2. A bacterial infection causes the number of white blood cells to increase. What is the *most* likely explanation?

   (1) More phagocytes are needed to fight the infection.
   (2) Red blood cells are decreasing.
   (3) Bacteria produce white blood cells.
   (4) Antibodies produce white blood cells.
   (5) Mucus is producing antibodies.

   Answer: (1) According to the article, a phagocyte attacks bacteria in the body. Therefore, during a bacterial infection more phagocytes than normal would be necessary to fight the infection.

READING: EVALUATION

Practice

Be aware that your own values and beliefs may affect how you answer a test question. Remember that the correct answer will reflect the author's point of view.

DIRECTIONS: Choose the <u>one</u> best answer for each item below.

Items 1–4 refer to the following passage.

Bacteria are small one-celled organisms that live almost everywhere. The largest bacteria are only about 1/65,000 centimeters across. Many are seven times smaller than that. Because they are so small, one of the ways of identifying bacteria is by their shape. Circular bacteria are called **cocci**. Rod-shaped bacteria are called **bacilli**. Coiled or corkscrew-shaped bacteria are called **spirilla**.

Most bacteria are enclosed in a capsule. The capsule protects the bacteria from its environment. Next to the capsule is a rigid cell wall composed of sugars and chains of amino acid molecules. Next to the cell wall is a cell membrane that allows certain substances to pass into and out of the bacterium itself. Bacteria have a nuclear area made up of a single chromosome of DNA, but no nuclear membrane surrounds it. Genetic material may also be found in small structures called **plasmids**. The whip-like action of structures called **flagella** allows some bacteria to move through their environment.

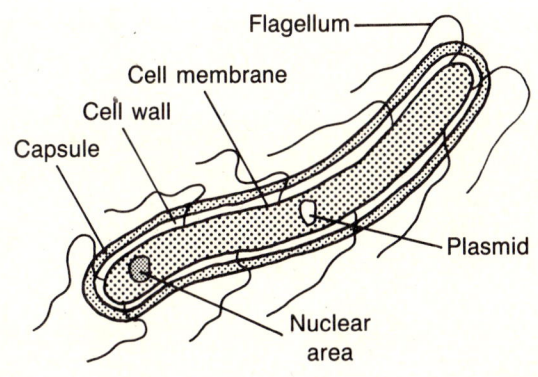

Many bacteria are helpful. Some break down decaying organic matter so it can be used by living plants and animals. People use bacteria to make cheese, yogurt, vinegar and wine.

Some bacteria are harmful. Many cause diseases such as pneumonia and tuberculosis. Certain bacteria in food can cause severe poisoning such as botulism or salmonellosis.

1. What is the shape of the bacteria *Diplococcus pneumoniae* which causes pneumonia?

    (1) rod
    (2) circular
    (3) coiled
    (4) corkscrew
    (5) flagella

2. What cell structure protects bacteria that live inside other organisms?

    (1) cell membrane
    (2) cell wall
    (3) capsule
    (4) plasmid
    (5) ribosome

GO ON TO THE NEXT PAGE.

3. One type of streptococcus, which does not have a capsule, is destroyed by the body. Which statement explains why?

   (1) It is made of sugars and amino acids.
   (2) It is made of cellulose and sugars.
   (3) It has no protection from the environment.
   (4) Flagella allow movement in the body.
   (5) It is composed of DNA.

4. The bacterium *Clostridium botulinum* causes which of the following kinds of food poisoning?

   (1) salmonellosis
   (2) tetanus
   (3) pneumonia
   (4) diphtheria
   (5) botulism

Items 5–8 refer to the following passage.

A virus is made up of a single molecule of genetic material—either DNA or RNA, but not both—surrounded by a coat of protein called a **capsid.** The capsid protects the genetic material of the virus from substances that could break it down.

A virus shows no signs of life as long as it is outside a living cell. It does not grow, reproduce or perform any chemical activity necessary for life. In order to become active, a virus must invade a living cell called a **host cell.** Once inside the host cell, the virus transfers its own genetic material into the cell and takes over all of the cell's activities. The host cell is then used to duplicate the virus' own genetic material and viral protein. These products are combined into new virus particles called **virions.** Each virion is exactly like the parent virus and can leave the host cell to infect other cells and begin the cycle over again.

Most viruses can multiply in only certain kinds of cells. Some infect specific plants like potato or tobacco plants. Viruses that attack animals often attack specific tissues. Cold viruses, for example, attack the tissues that line the nose and throat. Polio viruses attack certain nerve cells in the brain and spinal cord.

5. If a virus' capsid were destroyed, what would be the *most* likely outcome?

   (1) The virus would reproduce.
   (2) The virus would invade a host cell.
   (3) The viral genetic material would break down.
   (4) The virus would become active.
   (5) Virions would invade a host cell.

6. A virus takes over the activities of its host cell by

   (1) transferring its own genetic material into the host cell
   (2) reproducing itself in the host cell
   (3) releasing virions into the host cell
   (4) releasing its capsid into the host cell
   (5) breaking down the cell's membrane

7. If a host cell contains virions, and the cell dies, the virions would *most* likely

   (1) die
   (2) be released to invade other cells
   (3) use the host cell for food
   (4) take the place of the host cell
   (5) develop capsins

8. Scientists knew about viruses before they saw them. This was possible because scientists had

   (1) identified the viral nucleus
   (2) seen viral capsins
   (3) seen virions
   (4) identified host cells
   (5) recognized diseases viruses caused

GED Mini-Test 6

TIP When reading test passages, charts, diagrams or graphs, try to connect the new material to something you already know. In this way you can improve your memory and, therefore, be more successful on the GED test.

DIRECTIONS: Choose the one best answer for each item below.

Items 1–6 refer to the following passage.

Scientists have developed several ways in which to combat infectious diseases. Among them are vaccines, antibiotics and chemotherapy.

Vaccines Vaccines are substances that are made from killed or weakened bacteria or viruses and their toxins. When a particular vaccine is introduced into the body, the immune system develops antibodies to the disease. Antibodies are substances that make viruses and bacteria ineffective. Vaccines must be strong enough to trigger the production of antibodies, but weak enough so they will not cause the disease. Some vaccines provide protection for a long period of time. Others are effective for only a short period of time and additional amounts of vaccine are necessary at regular intervals. Some of the diseases that effective vaccines have been developed against include diphtheria, German measles, influenza, measles, mumps, whooping cough, polio, rabies, smallpox, tetanus and yellow fever.

Antibiotics Antibiotics are chemicals produced by living organisms and used to treat many bacterial diseases. Most antibiotics come from bacteria or molds. The most common antibiotics are penicillin and a group known as tetracyclines. Antibiotics destroy bacteria by disrupting their cellular chemistry. A good antibiotic is one that affects a wide range of disease organisms without harming the person who is receiving the antibiotic. Some of the diseases for which there are effective antibiotics include pneumonia, typhoid fever, tuberculosis and scarlet fever.

Chemotherapy Chemotherapy is a method of treating diseases with drugs made synthetically from chemicals. These drugs kill disease organisms without harming the body. One group of drugs is made from sulfur compounds. Sulfa drugs have been used to treat pneumonia, dysentery and meningitis. In some cases, scientists have been able to change the chemical structure of naturally occurring drugs to make them better.

1. Which group of disease-fighting substances comes from nature?

    (1) vaccines
    (2) antibiotics
    (3) chemotherapy drugs
    (4) sulfa drugs
    (5) tetracyclines

2. Which of the following would be given to a person to prevent the occurrence of a disease?

    (1) antibiotics
    (2) tetracyclines
    (3) chemotherapy
    (4) toxins
    (5) vaccines

GO ON TO THE NEXT PAGE.

3. If a person was exposed to measles but did not catch them, which of the following is *probably* true?

   (1) exposure not long enough
   (2) has antibodies against measles
   (3) has antibiotics against measles
   (4) already has measles
   (5) has antimeasles bacteria

4. Which of the following would be *most* effective in treating tuberculosis?

   (1) tetanus
   (2) sulfa
   (3) vaccine
   (4) toxin
   (5) penicillin

5. Polio was once epidemic in the United States. What was probably *most* responsible for stopping the spread of this viral disease?

   (1) antibiotics
   (2) sulfa drugs
   (3) tetracyclines
   (4) a vaccine
   (5) chemotherapy

Check your answers to the GED Mini-Test on page 55.

Answers and Explanations

Practice pp. 51–52

1. **Answer:** (2) The name of the species of bacteria, *Diplococcus pneumoniae*, tells you that the shape of the bacteria is coccal, or circular. Choice (1) refers to bacilli bacteria. Choices (3) and (4), coiled or corkscrew-shaped bacteria, are called spirilla. Choice (5) is the name of a structure found on some bacteria.

2. **Answer:** (3) The capsule protects a bacterium from its environment. Therefore, bacteria living inside other organisms would be protected from that chemical environment by the capsule as well. Choices (1), (2), (4) and (5) name other bacterial structures.

3. **Answer:** (3) Since the capsule protects the bacteria from its environment, the streptococcus bacteria without a capsule is unprotected and prone to destruction. Choice (1) refers to the composition of the cell wall, not capsule. Choice (2) is incorrect because there is no cellulose in bacteria. Choices (4) and (5) are irrelevant since they have nothing to do with the capsule.

4. **Answer:** (5) Botulism is one of the two types of food poisoning mentioned in the article. It is clear that it takes its name from the bacterium that causes it, *Clostridium botulinum*. Choice (1) names another kind of food poisoning. Choices (2) through (4) name diseases caused by bacterial infections.

5. **Answer:** (3) Since the capsid protects the genetic material, if the capsid was destroyed the genetic material could break down. Choice (1) and (4) are irrelevant since neither can happen until a virus invades a host cell. Choice (2) would be impossible once the virus had lost it capsid. Choice (5) is irrelevant since the question is talking about a virus and not virions.

6. **Answer:** (1) A virus takes over the activities of a host cell by transferring its genetic material into the cell. Choices (2), (3) and (4) cannot happen until the virus has already taken over the cell's activities. Choice (5) is irrelevant since breaking down the cell's membrane would kill the cell.

7. **Answer:** (2) The most likely answer to the question is that the virions would be released to invade other cells. There is no reason to suspect they would die, choice (1). Virions, being viruses, do not take in food, choice (3). Virions could not accomplish choice (4) since they cannot function by themselves. Virions already have choice (5).

8. **Answer:** (5) The only possible way scientists could have known about viruses before they saw them was by observing their effect on living things, namely the diseases they caused. Choices (1) through (3) could not occur unless scientists could already see viruses. Without seeing viruses choice (4) would be unknown.

GED Mini-Test pp. 53–54

1. **Answer:** (2) Since antibiotics come from living organisms such as molds and bacteria, it can be said that they come from nature. Choice (1) comes close, but is not found in its useable form in nature. Choices (3) through (5) refer to synthetically made substances.

2. **Answer:** (5) Vaccines are given to prevent diseases from ever happening. Choices (1) through (3) relate to substances that are used for curing diseases. Choice (4) is something that can cause a disease.

3. **Answer:** (2) Since measles is an infectious disease, if a person were exposed to it and did not catch it, the most probable reason would be that he or she had antibodies to the disease. Exposure, choice (1), is vague and not, therefore, the most probable answer. Choice (3), antibiotics, is a treatment for the disease, not a prevention. Choice (4) is irrelevant in regard to the question. Choice (5) is irrelevant since no such bacteria exists.

4. **Answer:** (5) Antibiotics are effective against bacterial infections such as tuberculosis. Penicillin is the only antibiotic mentioned. Choice (1) is the name of a disease. Sulfa, choice (2), is not an antibiotic. Choices (3) and (4) are not antibiotics, and are useful only in preventing disease, not in treating it.

5. **Answer:** (4) Stopping the spread of polio, or any other viral disease, is best accomplished by developing and administering a vaccine. Choice (1) is a treatment for bacterial diseases, and the question states that polio is a viral disease. Choices (2), (3) and (5) may all have been used to treat polio and may have helped to slow the spread of the disease. However, the question asks which was probably the *most* responsible for stopping the disease.

7 Mendelian Genetics

How often have you heard someone say of a newborn, "She looks just like her father"? Or "He has his mother's eyes and his father's chin"? Such statements refer to **traits,** or characteristics, we inherit from our parents. The study of inherited characteristics is called **genetics.**

The person most often credited with being the father of modern genetics was a monk named Gregor Mendel. Mendel began his experiments in Austria in 1857. He used pea plants in his experiments because of their many easily identifiable characteristics.

Mendel began his experiments by developing **purebred** strains of the pea plant. These were strains that always produced the same traits generation after generation. Next, he **crossed,** or bred, plants with contrasting traits. For example, he crossed purebred strains of tall plants with purebred strains of dwarf plants. The offspring of this cross are called **hybrids** because they contain different genes for the same trait, in this case, height. Mendel called this generation the F_1 **generation,** and, much to his surprise, all of the plants in the F_1 generation turned out to be tall plants!

Next, Mendel crossed tall hybrid plants from the F_1 generation. The offspring from this cross are called the F_2 **generation.** Mendel found that some of the plants in the F_2 generation were tall and some were dwarf. The number of tall and dwarf plants was always in a ratio of 3 to 1; that is, there was 1 dwarf plant for every 3 tall plants.

To explain his results, Mendel reasoned that since the offspring inherited genes from both parents, some traits were more powerful than other traits. He called these traits **dominant.** The traits that did not show up he called **recessive.** In the F_2 generation, the plants that show the recessive trait for dwarfness must, therefore, be purebreds that have no dominant genes for tallness. An illustration of Mendel's results is shown in the diagram below.

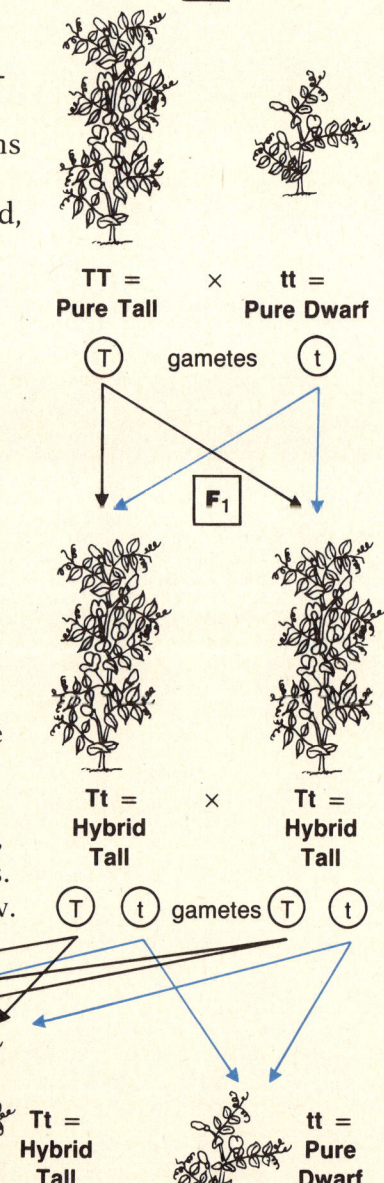

Strategies for READING

Assess the Appropriateness of Data to Prove or Disprove Statements

This skill involves applying what you know about fact and opinion, logical fallacies and unstated assumptions to determine how accurate a given statement is.

In order to determine if a **statement**, conclusion or answer is accurate, use the following questions to evaluate the **data:** Is the data based on fact or opinion? Does the statement or conclusion follow logically from the data that is presented? Is the data based on unstated assumptions? If so, what are they and are they logical?

Read the fifth paragraph in the article on page 56. By assessing what he already knew and comparing it to the data he observed, Mendel concluded that some traits were recessive and some were dominant. Was the data Mendel assessed appropriate? Was there any logical fallacy in Mendel's assessment?

Examples

DIRECTIONS: Use the information on this and the preceding page to choose the one best answer for each item below.

1. Mendel developed a strain of pea plants that produced *only* yellow seeds. These plants were

   (1) hybrids
   (2) recessive traits
   (3) dominant traits
   (4) purebreds
   (5) contrasting traits

Answer: (4) By assessing the data in the article, you can conclude that a plant that always produces the same trait is a purebred. Choice (1) is a plant that contains genes for more than one contrasting trait. Choices (2), (3) and (5) refer to traits, not strains, of plants.

2. If you cross pea plants with inflated pods with pea plants with constricted pods, all the F_1 generation will have inflated pods. Which of the following conclusions can you draw?

   (1) constricted pods are dominant
   (2) inflated pods are recessive
   (3) inflated pods are dominant
   (4) constricted pods are hybrid
   (5) inflated pods are hybrid

Answer: (3) By evaluating Mendel's data, you would conclude that inflated pods are a dominant trait. Choices (1) and (2) state the opposite, while choices (4) and (5) do not state a conclusion you could draw by observing the plants.

READING: EVALUATION

Practice

Evaluate the appropriateness of the data you use to answer questions. Your assessment of the data may mean the difference between a correct or an incorrect answer.

DIRECTIONS: Choose the one best answer for each item below.

Items 1–6 refer to the following passage and diagram.

A **Punnett square** can be used to predict the outcome of a genetic cross. Letters representing the genes from the male parent are placed across the top of the square. Letters representing the genes from the female parent are placed down the lefthand side of the square. The diagram shows a Punnett square used to identify the outcome of crossing pea plants with two sets of contrasting traits—short and dwarf plants; red and white flowers.

Each square is filled in by combining the genes at the top of the column with the genes at the beginning of the row. The results show the actual genes in the cells of each individual. This is called the **genotype.** The dominant trait is indicated with a capital letter; the recessive trait is indicated with a lowercase letter. In the diagram at right, T stands for tall, t stands for dwarf, R stands for red and r stands for white. Tall and red are dominant traits.

How the individual will actually look is called the **phenotype.** By identifying the dominant traits, you can determine the phenotype of any individual. For example, the phenotype of an individual with a genotype of TtRr will be tall with red flowers. An individual that has no contrasting genes will have the phenotype indicated by the genes it has. For example, an individual with the genotype ttRr will be dwarf with red flowers. An individual with a genotype of ttrr will be dwarf with white flowers.

	Male Genes			
	TR	tR	Tr	tr
TR	TTRR	TtRR	TTRr	TtRr
tR	TtRR	ttRR	TtRr	ttRr
Tr	TTRr	TtRr	TTrr	Ttrr
tr	TtRr	ttRr	Ttrr	ttrr

(Female Genes on left axis)

T = tall, dominant   t = dwarf, recessive
R = red, dominant   r = white, recessive

1. Which of the following shows a phenotype for being tall with white flowers?

   (1) ttrr
   (2) TtRr
   (3) Ttrr
   (4) ttRr
   (5) TtRR

2. Which of the following genotypes is made up of *only* recessive genes?

   (1) TTRR
   (2) TTrr
   (3) TtRr
   (4) ttRR
   (5) ttrr

GO ON TO THE NEXT PAGE.

3. How many different phenotypes are shown in the diagram?

(1) 5
(2) 4
(3) 3
(4) 2
(5) 1

4. According to the diagram, how many phenotypes will be tall with red flowers?

(1) 9
(2) 6
(3) 4
(4) 3
(5) 1

5. What is the number of *different* genotypes represented in the diagram?

(1) 16
(2) 10
(3) 9
(4) 5
(5) 3

6. What will a plant from a seed with the genotype of ttRr look like?

(1) tall with red flowers
(2) dwarf with white flowers
(3) tall with red flowers
(4) tall with pink flowers
(5) dwarf with red flowers

Items 7–10 refer to the following paragraph.

When both genes of a pair are identical in an individual, the individual is **homozygous** for the trait controlled by the genes. For example, a pea plant that has two R genes for red flowers is homozygous for that trait. An organism may be homozygous dominant or homozygous recessive depending on whether the pair of genes is for a dominant or recessive trait. When a pair of genes is not the same—for example, when one gene is for red flowers (R) and the other for white flowers (r)—the organism is **heterozygous** for the trait. Heterozygous organisms are also called hybrids.

7. If a trait that is not evident in the parents appears in their offspring, the parent genotypes are *most* likely

(1) pure recessive
(2) monoploid
(3) homozygous
(4) heterozygous
(5) diploid

8. Which of the following pair of traits would indicate that an individual is heterozygous?

(1) recessive white/recessive white
(2) dominant red/dominant red
(3) dominant tall/dominant tall
(4) recessive dwarf/recessive dwarf
(5) dominant tall/recessive dwarf

9. Which of the following gene notations would indicate that an individual is homozygous for two traits?

(1) TtRR
(2) TTRr
(3) TTrr
(4) TrRr
(5) TT

10. A pair of black (B) mice produce some offspring that are black and some that are white (b). The genotype of the parents is *most* probably

(1) BB and bb
(2) BB and Bb
(3) Bb and Bb
(4) bb and bb
(5) BB and Bb

Before you take the GED Mini-Test, check your answers on pages 61–62.

GED Mini-Test

7 TIP Analyze the types of test questions so you are *prepared* for a test. Skim the test before actually taking it so you know what you are being expected to do. This will also help you manage your test-taking time.

DIRECTIONS: Choose the one best answer for each item below.

Items 1–6 refer to the following passage and diagram.

Chromosomes are actually made up of genes, or DNA molecules. The DNA molecule itself is made up of units called nucleotides. Look at the diagram to the right showing a nucleotide of a DNA molecule. A nucleotide consists of: (1) a sugar, (2) a phosphate and (3) a nitrogen base. Many nucleotides link together in such a way that the DNA molecule looks like a ladder. The upright sides of the "ladder" contain the phosphate and sugar of the nucleotide. The "rungs" are the nitrogen bases.

There are four different bases on the rungs; these bases are identified as A, G, T and C. A always links with T, and G always links with C, so they are called complementary pairs. Each "rung" of the DNA ladder consists of the following pairs of bases: AT, GC, CG and TA, and their arrangement determines the code of each DNA molecule or gene.

When a cell divides, the chromosomes, or DNA molecules, make exact copies of themselves. The diagrams below follow the steps involved in the formation of a new DNA molecule. Figure 1 shows a model of the DNA molecule. When DNA duplication occurs, the DNA molecule is separated into two halves (figure 2). The hydrogen bonds that join the nucleotide bases are broken by enzymes and energy from the cell. Then free nucleotides in the cell move in to find their complementary bases (figure 3). As a result, two new DNA molecules that are exactly like the original (figure 4) are formed.

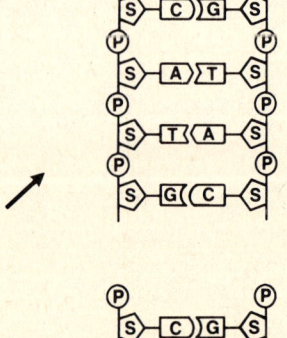

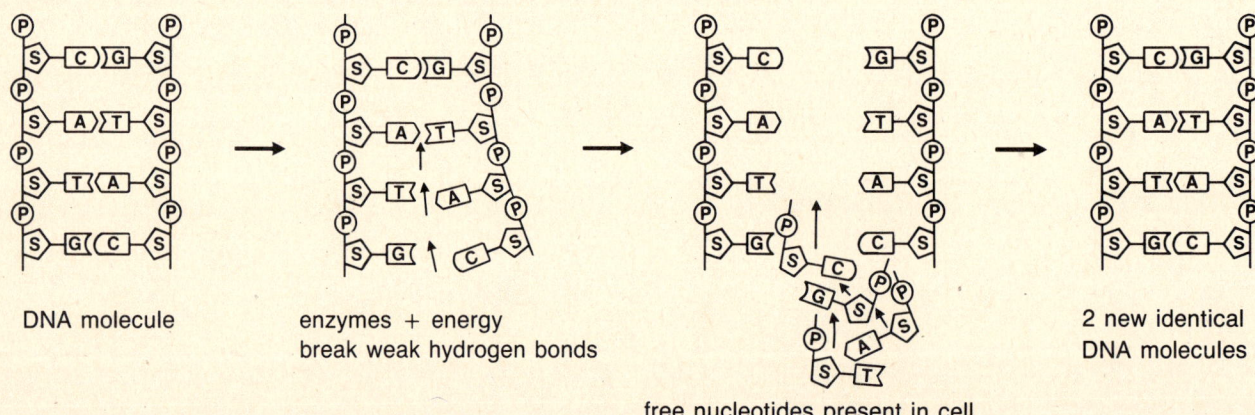

GO ON TO THE NEXT PAGE.

1. How many sugar molecules are attached to *each* nucleotide?

   (1) 5
   (2) 4
   (3) 3
   (4) 2
   (5) 1

2. Which of the following pairs of nucleotide bases is complementary?

   (1) cytosine (C)/thymine (T)
   (2) thymine (T)/guanine (G)
   (3) guanine (G)/cytosine (C)
   (4) guanine (G)/thymine (T)
   (5) adenine (A)/guanine (G)

3. Adenine (A), sugar and a phosphate make up a

   (1) nucleotide
   (2) DNA molecule
   (3) chromosome
   (4) hydrogen bond
   (5) complementary pair

4. Guanine (G) and cytosine (C) are held together in a DNA molecule by

   (1) adenine
   (2) nucleotides
   (3) enzymes
   (4) phosphate
   (5) hydrogen bonds

5. What would be the complementary sequence to a nucleotide base sequence of GTCA?

   (1) GTCA
   (2) ACTG
   (3) CAGT
   (4) TGAC
   (5) AGCT

6. If there were no free nucleotides in a cell, which would *most* likely be the result?

   (1) The cell would divide.
   (2) The cell would not be able to pass on hereditary information.
   (3) The chromosomes would make nucleotides.
   (4) The cell would begin to pass on hereditary information.
   (5) The chromosomes would duplicate.

   Check your answers to the GED Mini-Test on page 62.

Answers and Explanations

Practice pp. 58–59

1. **Answer:** (3)   The traits you actually see make up the phenotype. The genotype Ttrr shows a dominant gene for tallness and a pair of recessive genes for white flowers. You would, therefore, see a tall plant with white flowers. Each of the other alternatives describes a different kind of phenotype: choice (1) dwarf/white, choice (2) tall/red, choice (4) dwarf/red and choice (5) tall/red.

2. **Answer:** (5)   Recessive genes are indicated by lowercase letters. A genotype made up only of recessive genes would, therefore, be indicated by all lowercase letters. The only choice that fits the criterion is ttrr. Choice (1) has all dominant genes. Choices (2) through (4) each show two dominant genes and two recessive genes.

3. **Answer:** (2)   The four different phenotypes shown in the diagram are: tall/red, dwarf/red, tall/white and dwarf/white. These phenotypes, however, may have different genotypes.

4. **Answer:** (1)   Nine of the 16 genotypes will produce tall plants with red flowers. They are: one with the genotype TTRR, two with the genotype TtRR, two with the genotype TTRr and four with the genotype TtRr.

5. **Answer:** (3)   The key word in the question is *different*. Each set of genes represents a genotype, but there are only nine different genotypes. These are: TTRR, TtRR, TTRr, TtRr, ttRR, ttRr, TTrr, Ttrr and ttrr.

6. **Answer:** (5)   The genotype ttRr is pure for the dwarf trait (tt) and dominant for red flowers (Rr). All the other choices have a different phenotype than dwarf with red flowers.

7. **Answer:** (4) The parents are most likely heterozygous and carry recessive genes that combine and appear in their children. If the parents were homozygous, choice (3), or pure recessive, choice (1), the children would be the same as the parents. Monoploid, choice (2), and diploid, choice (5), were not mentioned in the paragraph.

8. **Answer:** (5) Heterozygous individuals carry a pair of contrasting genes for the same trait. In this case, the trait is height and the contrast, or variation, is tall and dwarf. Choices (1) through (4) all name genetic traits that are the same or homozygous.

9. **Answer:** (3) Homozygous means that both genes for a trait are the same. TTrr shows that the pair of genes for height (T) are the same, and the pair of genes for color (r) are the same. Choices (1) and (2) have only one pair of genes that are the same. Choice (4) shows that neither pair of genes is the same. Choice (5) show a pair of genes for only one trait.

10. **Answer:** (3) Since some of the offspring are black and some are white, the parents must carry a recessive gene for white for it to appear. Choices (1), (2) and (5) are incorrect since they would produce all black offspring. Choice (4) would produce all white offspring.

GED Mini-Test *pp. 60–61*

1. **Answer:** (5) By looking at the diagram, it is clear that each nucleotide has only one sugar molecule.

2. **Answer:** (3) Complementary nucleotides are nucleotides whose bases link together. Guanine (G) and cytosine (C) always link to one another. None of the other pairs of nucleotide bases listed as choices will link together.

3. **Answer:** (1) A nucleotide is made up of a base, sugar and phosphate. One of the bases is adenine (A); therefore, adenine, sugar and phosphate make up a nucleotide. Choices (2) and (3) are made up of nucleotides. Choice (4) is irrelevant. Choice (5) refers to two nucleotide bases that link together.

4. **Answer:** (5) (G) and (C) are two complementary nucleotide bases. Nucleotide bases are held together by hydrogen bonds. Choice (1) names another nucleotide base. (G) and (C) are part of choice (2). Choice (3) helps to break apart nucleotide bases. Choice (4) is part of a nucleotide.

5. **Answer:** (3) (G) always bonds to (C) and (T) always bonds to (A). Therefore, the sequence GTCA would bond with the sequence CAGT. Each of the other choices names a sequence that could not occur.

6. **Answer:** (2) Remember that a chromosome passes on hereditary information and is made up of a DNA molecule. In order for a DNA molecule to duplicate, it must have free nucleotides from the cell to bond with its own nucleotide bases. If the cell could not provide nucleotides, the DNA molecule, or chromosome, could not pass on its hereditary information. Choice (1) is unlikely to happen. A chromosome is made up of nucleotides, choice (3), but does not make them. Choices (4) and (5) are the opposite of the correct answer.

8 Ecosystems

An **ecosystem** is a natural community in which all the living and nonliving things in the community interact. Examples of ecosystems include lakes, rivers, meadows, ponds and swamps. The most important relationships in an ecosystem involve the flow of food and energy.

Food and the ecosystem. The movement of food through an ecosystem is called a **food chain.** A food chain begins with the producers, which in most ecosystems are green plants. Animals that eat the plants are called **primary consumers.** The animals that eat the primary consumers are called **secondary consumers.** In some ecosystems there may be **tertiary consumers** that eat the secondary consumers.

When the plants and animals in an ecosystem die, bacteria and fungi in the soil break them down into compounds that can be used by the plants. In this way the matter in an ecosystem is constantly being cycled through it. The organisms that break down decaying plants and animals are called **decomposers.**

Energy and the ecosystem. The sun is the main source of energy for an ecosystem. The sun's energy enters the system through green plants, which are able to store some of it. This energy, in turn, is passed along the food chain as consumers eat producers and other consumers. However, since living things need to use energy in order to carry out their life processes, as energy moves along the food chain, less and less of it is available. For example, only ten units of energy out of every 1,000 units is transferred from plants to primary consumers. Only about one unit out of every ten is transferred to secondary consumers from primary consumers. As you can see, the further along the food chain you go, less energy is available.

An ecosystem is a complex community of living things interacting with each other and their environment. An ecosystem is maintained by the flow of food and energy through it.

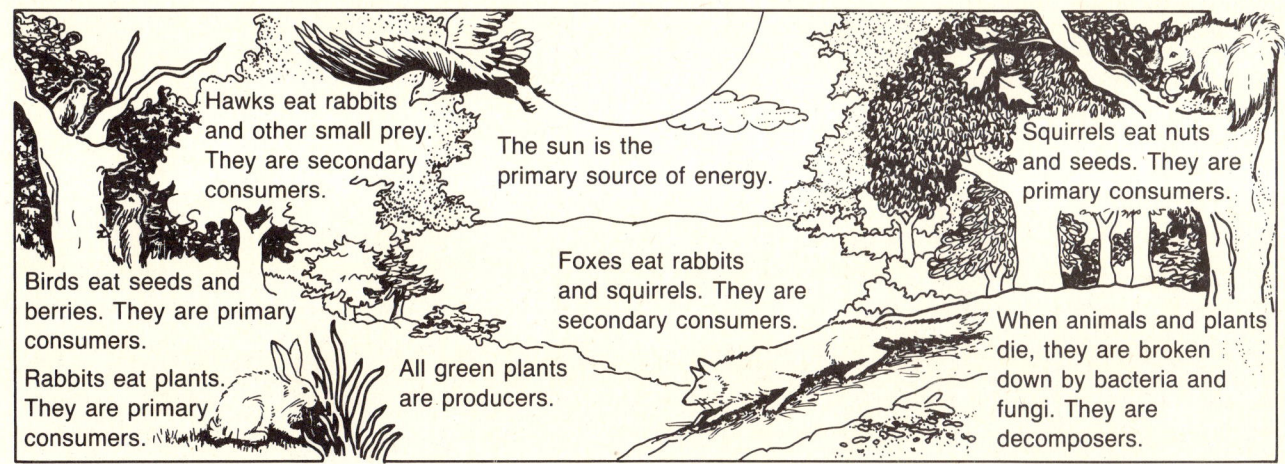

Hawks eat rabbits and other small prey. They are secondary consumers.

The sun is the primary source of energy.

Squirrels eat nuts and seeds. They are primary consumers.

Birds eat seeds and berries. They are primary consumers.

Foxes eat rabbits and squirrels. They are secondary consumers.

Rabbits eat plants. They are primary consumers.

All green plants are producers.

When animals and plants die, they are broken down by bacteria and fungi. They are decomposers.

INTRODUCTION

Strategies for READING

Identify Logical Fallacies in Arguments

This skill involves understanding and finding faulty logic and irrelevant evidence to prove or disprove statements.

In a **logical fallacy,** the argument presented to support a conclusion is based on faulty logic. For example, the conclusion drawn from the following argument is incorrect. Can you identify the logical fallacy? The article on page 63 holds the key.

> Decomposers in an ecosystem break down decaying matter into compounds that can be used by plants. In this way the matter is constantly being recycled. Therefore, decomposers are the most important part of the ecosystem.

The logic is false because it fails to recognize the cyclical nature of an ecosystem. The plants, animals and decomposers in an ecosystem all interact with one another. One element, therefore, is no more important than another.

Examples

DIRECTIONS: Use the information on this and the preceding page to choose the <u>one</u> best answer for each item below.

1. Why are squirrels labeled as primary consumers in the food chain?

   (1) They are decomposers.
   (2) They eat nuts and seeds.
   (3) They eat snakes.
   (4) They eat rabbits.
   (5) They are producers.

Answer: (2) Animals that eat plants are called primary consumers. Nuts and seeds are the products of plants. Decomposers, choice (1), break down decaying plant and animal matter. Choices (3) and (4) are animals that eat other animals. Choice (5) are the plants themselves.

2. All living things use energy to carry out life activities. Therefore, the source of the energy in an ecosystem *must* be

   (1) constantly recycled
   (2) in short supply
   (3) in abundant supply
   (4) in the primary producers
   (5) constantly resupplied by the sun

Answer: (5) While it is true that energy must be in the primary producers, choice (4), this does not complete the statement about the source of the energy (logical fallacy). Since energy is being used up, it cannot be recycled, choice (1), and would not be in abundant supply, choice (3). If energy were in short supply, choice (2), the system would not function long.

64   READING: EVALUATION

Practice

The best answer requires careful interpretation. Be sure your answer is based on a sound conclusion and not a logical fallacy.

DIRECTIONS: Choose the one best answer for each item below.

Items 1–4 refer to the following passage and bar graph.

Early people were **nomadic.** They wandered from area to area hunting animals and gathering plants for food. With the advent of agriculture, this nomadic existence gradually stopped, and people began to live in villages. The same amount of land could support many more people and the world's population began to increase. Today the world's birth rate is three times higher than the death rate. The obvious result is that the world's population will continue to rise unless something is done to stop it.

The increasing world population is a serious problem. Overpopulated areas of the world already suffer from disease, poverty and crime. Added to these problems are the limited supplies of such natural resources as water, food, wood and fossil fuels.

It is easier to understand the dramatic growth in world population by looking at a graph like the one at the right. In the year 10,000 B.C. it is estimated that at the most there were about 10 million people in the world. By the first century A.D. that number had grown to about 300 million, a thirtyfold increase in 10,000 years. The population remained fairly constant for the next 1,600 years. By 1650 the world's population had only risen to about 510 million people. During the next 200 years, however, the population more than doubled. Between 1850 and 1950 the population doubled again. The next doubling of the world's population only took 30 years, and another doubling is expected by the year 2015.

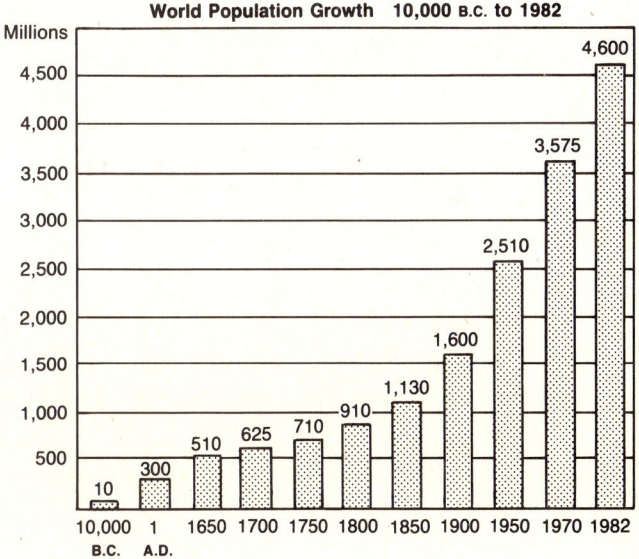

1. Which factor is probably *most* responsible for the increasing world population?

    (1) decrease in birth rate
    (2) increase in death rate
    (3) increase in a nomadic way of life
    (4) better weather conditions
    (5) increase in birth rate

2. The projected world population in the year 2015 is about

    (1) 12.5 billion
    (2) 9.2 billion
    (3) 7.0 billion
    (4) 6.25 billion
    (5) 5.6 billion

GO ON TO THE NEXT PAGE.

3. Which of the following is probably the *best* way to reduce increasing world population?

   (1) decreasing the death rate
   (2) increasing the birth rate
   (3) decreasing the birth rate
   (4) increasing the food supply
   (5) increasing the death rate

4. By about how much did the world population increase between 1900 and 1970?

   (1) 1.975 billion
   (2) 1.187 billion
   (3) 910 million
   (4) 770 million
   (5) 1.975 million

Items 5–8 refer to the following chart and paragraph.

World Population Density Per Square Kilometer

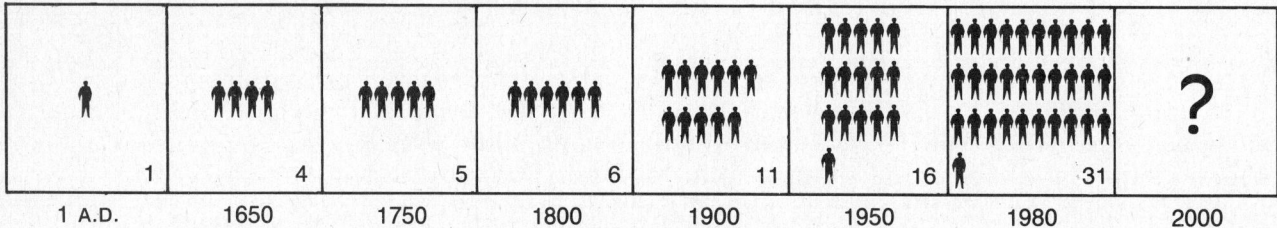

Human population density refers to the number of people that live in a specific area. Population densities are based on averages; that is, the amount of available land in an area is divided by the number of people living in the area. If an area's population increases, so does its population density. The chart above shows how the world's population density has changed from the year 1 A.D. to the year 1981.

5. How many more people occupied one square kilometer of space in 1980 than in 1950?

   (1) 31
   (2) 20
   (3) 16
   (4) 15
   (5) 5

6. About how many times greater was the world population density in 1980 than in 1900?

   (1) 10
   (2) 8
   (3) 5
   (4) 4
   (5) 3

7. According to the information in the graph, what is likely to happen to the world population density by the year 2000 if everything remains the same?

   (1) It is likely to triple.
   (2) It is likely to be diminished by one half.
   (3) It is likely to double.
   (4) It is likely to remain the same.
   (5) It is likely to go back to the 1980 density.

8. Which of the following accurately states the relationship between the size of a population and the population density for a given area?

   (1) The greater the population, the greater the population density.
   (2) The greater the population, the less the population density.
   (3) The less the population, the greater the population density.
   (4) Population density and the size of a population are the same.
   (5) The size of the population is inversely related to population density.

GED Mini-Test

TIP If taking notes helps you study, take notes when reading a test passage. Writing ideas in your own words makes them easier to understand, especially those that are complicated. Use your notes to help you answer the GED test questions.

DIRECTIONS: Choose the one best answer for each item below.

Items 1–6 refer to the following passage.

Without air, an important natural resource, most living things would die. Yet every day the smoke from factories and homes, and the exhaust from cars, trucks, airplanes and incinerators pours into the atmosphere. As a result, the air we breathe is often polluted.

Gases. When fossil fuels are burned, they release gases into the atmosphere. The most common of these gases include sulfur dioxide, nitric oxide and carbon monoxide. Sulfur dioxide is a heavy gas that results from burning coal and oil that contain sulfur. Sulfur dioxide combines with water to form sulfuric acid, which can be injurious to the eyes and linings of the lungs. When sulfur dioxide is inhaled, it usually causes choking, and in large doses it can be fatal.

Nitric oxide is a gas released when the engines of trucks and automobiles burn gasoline. Nitric oxide combines with oxygen in the air to form the deadly gas nitrogen dioxide. When nitrogen dioxide combines with the water in your eyes and lungs it forms nitric acid, which, in high concentrations, can cause permanent damage.

Carbon monoxide is a gas that is given off when organic substances such as coal, wood, oil and gasoline do not burn completely. The exhaust from automobiles is the major source of carbon monoxide. It combines with your blood much faster than oxygen does and can cause the body's tissues to die from lack of oxygen. In this way carbon monoxide can be fatal.

Acid rain. Rain normally has a slightly acid pH of from 5.6 to 7.0. A pH of 7.0 is neutral. Anything less is acidic and anything above it is alkaline. When sulfur and nitrogen dioxide are in the atmosphere, they react with rainwater to form sulfuric acid and nitric acid. These acids make the rainwater more acidic by lowering the pH. When the pH of rainwater is less than 5.6 it is called acid rain. Acid rain is very destructive to plant and aquatic life. In some parts of the world, entire forests and lakes have been killed by acid rain. In addition, acid rain eats away at metal and stone surfaces, often ruining statues and building façades.

Smog. When smoke and fog combine, the result is called smog. In large metropolitan areas smog can sometimes be seen as a gray- or rust-colored haze over the horizon. Smog in industrial cities is usually made up of smoke containing sulfur dioxide and fog. Automobile exhausts produce another kind of smog that is a combination of fog and harmful gases including nitrogen dioxide and carbon monoxide. Either kind of smog can be harmful to the lungs and eyes of people and animals.

GO ON TO THE NEXT PAGE.

1. Why do the eyes and lungs become irritated when it rains in coal-mining towns?

    (1) carbon dioxide combines with sulfur dioxide
    (2) water combines with sulfur dioxide
    (3) blood combines with sulfur dioxide
    (4) salt combines with sulfur dioxide
    (5) oxygen combines with sulfur dioxide

2. Which of the following pollutants would you expect to be produced by jet engines?

    (1) nitric acid
    (2) sulfur dioxide
    (3) smog
    (4) acid rain
    (5) nitric oxide

3. Which of the following is the *most* likely source of sulfur dioxide in smog?

    (1) factories
    (2) automobiles
    (3) acid rain
    (4) carbon monoxide
    (5) sulfuric acid

4. Acid rain is destructive to aquatic environments because it

    (1) raises the pH
    (2) makes the environment more alkaline
    (3) produces harmful gases
    (4) eats away at stones
    (5) lowers the pH

5. Charcoal is made from wood. Which pollutant is *most* likely to be the result of burning charcoal?

    (1) carbon monoxide
    (2) nitric oxide
    (3) nitrogen dioxide
    (4) sulfur dioxide
    (5) sulfuric acid

6. The amount of nitrogen dioxide in the atmosphere could be effectively reduced by lowering the

    (1) nitrogen dioxide emissions in cars
    (2) carbon monoxide emissions in cars
    (3) sulfur dioxide emissions in cars
    (4) nitric oxide emissions in cars
    (5) nitric acid emissions in cars

Check your answers to the GED Mini-Test on page 69.

Answers and Explanations

Practice *pp. 65–66*

1. **Answer:** (5) An increase in the birth rate is the greatest factor contributing to the increasing world population. Choices (1) and (2) would have the opposite effect. Choice (3) is not happening; in fact, very few nomadic peoples still exist today. Weather conditions, choice (4), have remained pretty much the same and cannot account for the increasing population.

2. **Answer:** (2) According to the article, the next doubling of the world's population will occur between 1982 and 2015. If the world population was about 4.6 billion people in 1982, double that figure would be about 9.2 billion people in the year 2015.

3. **Answer:** (3) Since an increasing birth rate is the major factor in population growth, the best way to decrease the rate of population growth would be to decrease the birth rate. Choices (1) and (2) would have the opposite effect; that is, they both would increase the population. Choice (4), even if possible in large enough quantities, would also contribute to increasing the population. Choice (5) is not acceptable in a civilized world.

4. **Answer:** (1) By subtracting the world population figure for 1900 from that for 1970, you can determine the increase in world population over that period of time. In order to arrive at the correct answer, you must have recognized that the figures in the graph are given in millions; that is, each figure above the bar must be multiplied by one million. Therefore, 1.600 is the same as 1,600,000,000 and 3.575 is the same as 3,575,000,000.

5. **Answer:** (4) According to the pictograph, 16 people occupied one square kilometer of space in 1950. In 1980, 31 people occupied one square kilometer of space. By subtracting the two figures, you can determine how many more people occupied the same amount of space in 1980 than in 1950.

6. **Answer:** (5) According to the pictograph, 11 people occupied one square kilometer of space in 1900. In 1980, 31 people occupied one square kilometer of space. Thirty-one is about three times more than 11.

7. **Answer:** (3) According to the rate of change shown in the pictograph, the population density doubled between 1900 and 1950, a period of 50 years. It doubled again between 1950 and 1980, a period of 30 years. It is reasonable to assume that the population will double again between 1980 and 2000, a period of 20 years.

8. **Answer:** (1) As world population increases, more people must occupy the same amount of land, since the amount of land is constant. Therefore, the greater the world population, the greater the population density. Choices (2) through (4) draw the wrong conclusion from the data. Choice (5) is not a logical statement and is irrelevant.

GED Mini-Test   pp. 67–68

1. **Answer:** (2) When sulfur dioxide combines with water it forms sulfuric acid. Therefore, when sulfur dioxide in the atmosphere of coal-mining towns combines with the water from rain, sulfuric acid is formed, and the eyes and lungs become severely irritated. The other choices are irrelevant since those substances do not combine with sulfur dioxide to form sulfuric acid.

2. **Answer:** (5) Engines produce the gas nitric oxide as a by-product of combustion. Choice (1) is the result of nitrogen dioxide combining with water. Choice (2) is produced as a result of burning fuels that contain sulfur. Choice (3) is a combination of gaseous pollutants and smoke in the atmosphere. Choice (4) is the result of acids in the atmosphere combining with rainwater.

3. **Answer:** (1) When fossil fuels that contain sulfur are burned, they give off sulfur dioxide. These fuels are used by many industrial factories. Therefore, you would expect the sulfur dioxide in smog to come from factories. Choice (2) is mainly responsible for nitric oxide. Choice (3) is a combination of acids in the atmosphere and rainwater. Choice (4) is the result of the incomplete burning of organic substances. Choice (5) results from sulfur dioxide combining with water.

4. **Answer:** (5) Acid rain causes bodies of water to become acidic. The more acid the aquatic environment is, the lower its pH level. Choices (1) and (2) are different ways of stating the opposite effect. Acid rain itself does not produce a gas, choice (3). Though choice (4) may be one of the results of acid rain, it is not the result that destroys the aquatic environment.

5. **Answer:** (1) Wood is an organic substance that produces carbon monoxide when it is burned incompletely. Therefore, charcoal, which is made from wood, is likely to give off carbon monoxide as well. Choice (2) is the result of burning fossil fuels in engines. Choice (3) is the result of nitric oxide combining with oxygen. Choice (4) is the result of burning fossil fuels that contain sulfur. Choice (5) is the result of sulfur dioxide combining with water.

6. **Answer:** (4) Exhaust from cars is one of the main sources of nitric oxide in the atmosphere. When nitric oxide combines with oxygen in the atmosphere, it produces nitrogen dioxide. Therefore, by reducing nitric oxide emissions in cars, the amount of nitrogen dioxide in the atmosphere should also be reduced. All other choices are incorrect because they name gases or acids that are not responsible for the formation of nitrogen dioxide.

ANSWERS AND EXPLANATIONS

REVIEW
Life Science

In this section you have read and studied about the structure of living organisms and about some of the chemical processes necessary for life. You have also studied how living things interact with one another and with their environment. You have learned how physical characteristics are passed from generation to generation and read about theories explaining the origin and evolution of life. The following exercises summarize and expand upon these concepts. Be aware of familiar words and ideas as you answer each set of questions. You will find that the more you read and study in the field of biology, the quicker you will be able to absorb new material as it is presented.

DIRECTIONS: Choose the one best answer for each item below.

Items 1–4 refer to the following diagram and definitions.

The diagram below shows the structure of a single-celled organism called a paramecium.

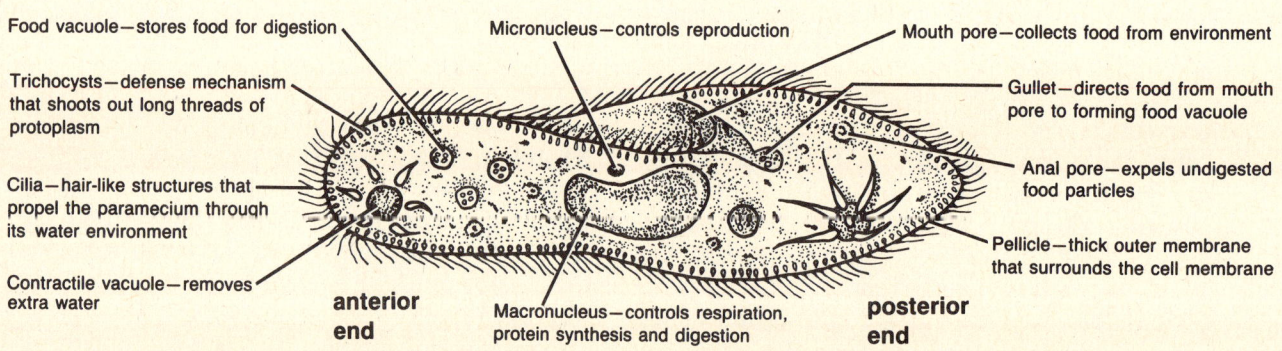

1. As a paramecium moves through its water environment, it picks up food particles. Into which structure do the food particles pass?

   (1) macronucleus
   (2) anal pore
   (3) gullet
   (4) contractile vacuole
   (5) trichocysts

2. A paramecium may reproduce by dividing into two new organisms. Which structure controls this process?

   (1) macronucleus
   (2) contractile vacuole
   (3) food vacuole
   (4) trichocyst
   (5) micronucleus

3. What would *most* likely happen if a paramecium did not have contractile vacuoles?

   (1) It would enlarge with water.
   (2) It would shrink.
   (3) It would starve.
   (4) It would not be able to get rid of waste.
   (5) It would not be able to reproduce.

4. According to the diagram, where are food vacuoles formed?

   (1) next to the macronucleus
   (2) in the contractile vacuoles
   (3) at the edge of the mouth pore
   (4) at the end of the gullet
   (5) in the pellicle

GO ON TO THE NEXT PAGE.

Items 5–10 refer to the following diagram and paragraph.

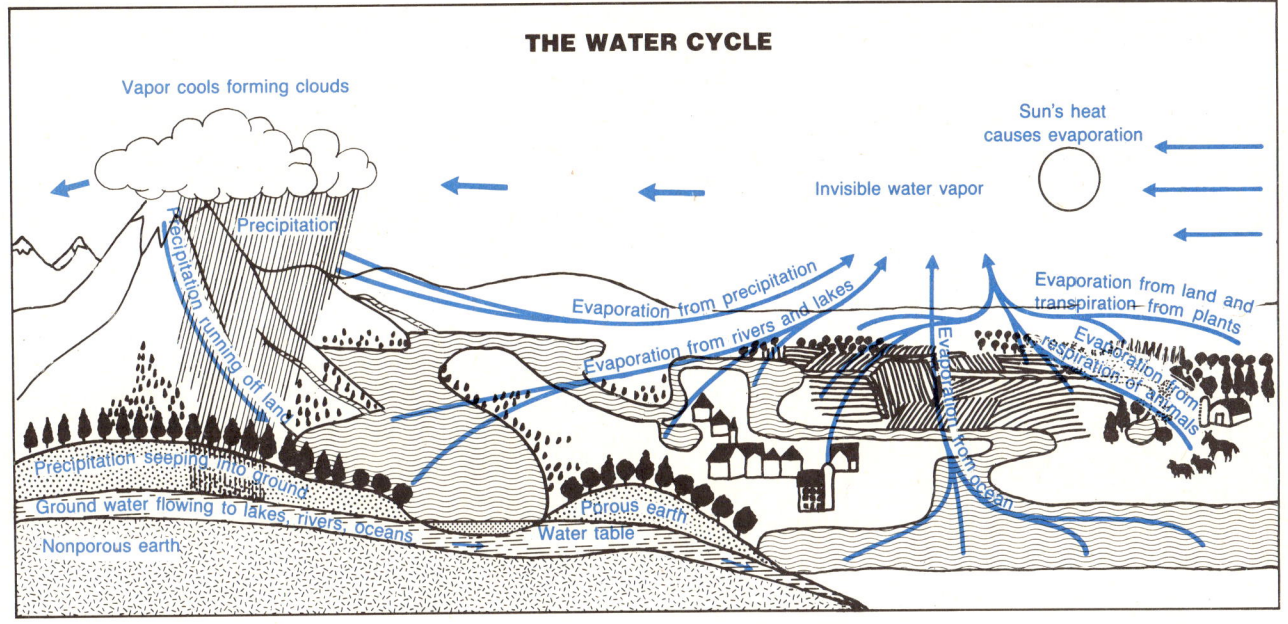

All living organisms must have water in order to live. Cells are composed largely of water, and water is required in order for the chemical reactions in living things to take place. However, the amount of the earth's water is limited. It must, therefore, be used again and again. The continual circulation of the earth's water is called the water cycle.

5. According to the diagram, the sun is responsible for

    (1) evaporation
    (2) respiration
    (3) transpiration
    (4) precipitation
    (5) ground water

6. The water that falls from clouds to the earth is called

    (1) transpiration
    (2) ground water
    (3) precipitation
    (4) evaporation
    (5) respiration

7. Water flows to the lakes, rivers and oceans in the form of

    (1) evaporation
    (2) precipitation
    (3) water table
    (4) water vapor
    (5) ground water

8. Which of the following is a source of atmospheric water?

    (1) precipitation
    (2) ground water
    (3) transpiration
    (4) water vapor
    (5) water table

9. Destroying vast forest lands could affect the amount of water in the atmosphere by

    (1) reducing the amount of water from transpiration
    (2) increasing the amount of water from transpiration
    (3) increasing the amount of water from evaporation
    (4) reducing the amount of water from evaporation
    (5) polluting lakes and rivers

10. One way the water animals use is returned to the atmosphere is through

    (1) transpiration
    (2) precipitation
    (3) evaporation from precipitation
    (4) respiration
    (5) photosynthesis

GO ON TO THE NEXT PAGE.

Items 11–16 refer to the following diagram.

GEOLOGIC TIME AND EVOLUTION

Era	Period	Record of Life Change	Millions of Years Ago
Cenozoic (age of mammals)	Quaternary	Humans dominate life; many organisms become extinct.	
	Tertiary	Anthropoid apes; many mammals and flowers present.	— 1
			— 70 —
Mesozoic (age of reptiles)	Cretaceous	Hardwood trees; dinosaurs begin to die out; marsupials present.	
			— 135 —
	Jurassic	Many dinosaurs; first mammals; first feathered birds.	
			— 180 —
	Triassic	Reptiles and turtles present; cone-bearing trees.	
			— 225 —
Paleozoic (age of amphibians) (age of fish) (age of invertebrates)	Permian / Carboniferous	Many amphibians; first reptiles; ferns; insects present.	
			— 270 —
	Devonian / Silurian	Many fish and sharks; first amphibians; first trees.	
			— 400 —
	Ordorvician / Cambrian	First land plants; first fish; many invertebrates present	
			— 600 —
Precambrian	Precambrian	Few fossils; some shells, algae, fungi, bacteria and soft-bodied invertebrates.	
			— 4,500

11. According to the chart, in which period did mammals first appear?

(1) Quaternary
(2) Cretaceous
(3) Jurassic
(4) Devonian
(5) Precambrian

12. According to the chart, which of the following kinds of animals evolved first?

(1) mammals
(2) reptiles
(3) amphibians
(4) birds
(5) fish

13. About how many years ago did the first human-like animals appear?

(1) 1 million
(2) 70 million
(3) 180 million
(4) 400 million
(5) 4.5 billion

14. Which of the following kinds of plants are the *most* recently evolved?

(1) ferns
(2) conifers
(3) hardwood trees
(4) flowering plants
(5) algae

15. According to the chart, life has evolved

(1) from the complex to the simple
(2) from the simple to the complex
(3) in a short period of time
(4) all at once
(5) through major catastrophes

16. Fossil evidence provides clues to when certain kinds of animals lived. If you discovered a dinosaur fossil, in which of the following eras would you place it?

(1) Precambrian   (2) Paleozoic
(3) Jurassic   (4) Mesozoic
(5) Cenozoic

GO ON TO THE NEXT PAGE.

Items 17–22 refer to the following diagram and paragraph.

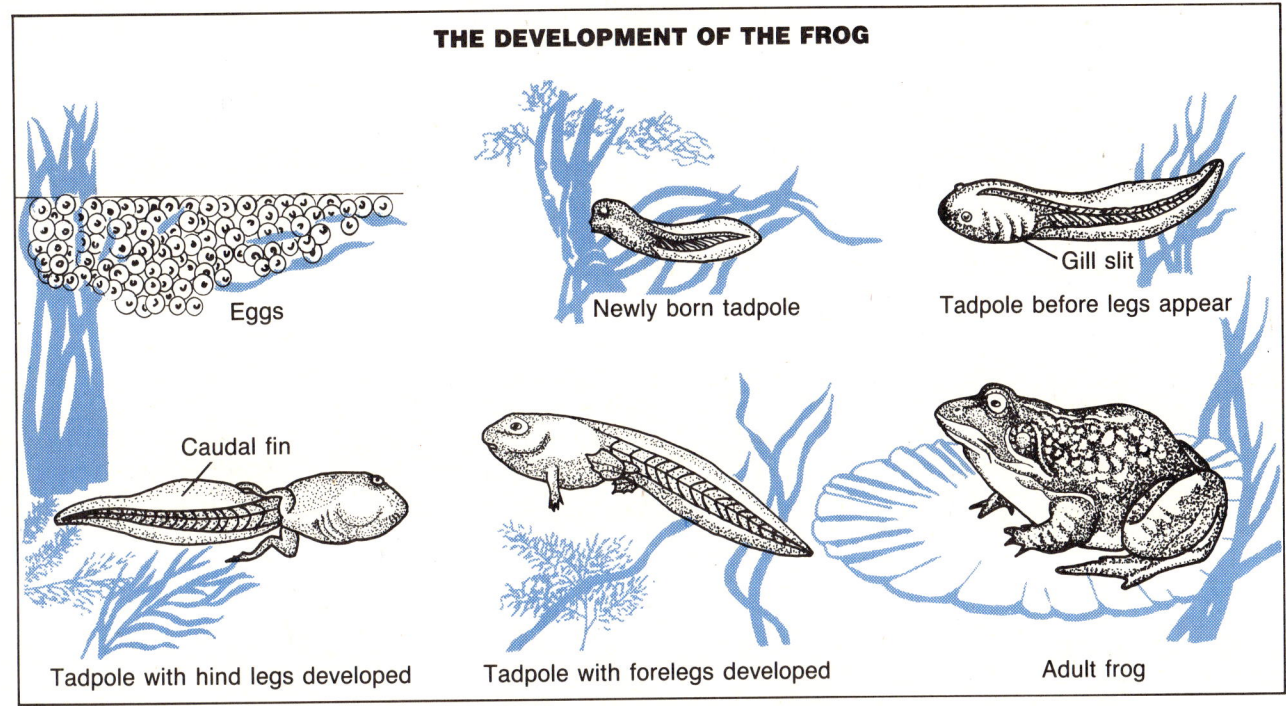

The life cycle of the frog demonstrates the process of metamorphosis. During metamorphosis, the larval, or immature, form of an organism gradually changes to a very different adult form. Metamorphosis enables frogs and other amphibians to develop from an aquatic larval form to a terrestrial, or land-dwelling, adult form.

17. The *most* obvious function of a tadpole's tail is to

(1) help it swim
(2) supply oxygen
(3) ward off enemies
(4) manufacture food
(5) keep the tadpole underwater

18. What is the *most* obvious difference between the mouth of an adult frog and that of a tadpole?

(1) It is much smaller and narrower.
(2) It is much larger and broader.
(3) A frog has a mouth; a tadpole does not.
(4) It is located in a different place.
(5) There is no difference.

19. It is necessary for a tadpole to have gills in order to

(1) obtain food
(2) excrete waste material
(3) reproduce
(4) breathe air
(5) breathe underwater

20. Which adult-like characteristic appears first in the tadpole?

(1) gills
(2) tail
(3) hind legs
(4) forelegs
(5) caudal fin

21. Which of the following *must* develop before a tadpole can become a land-dwelling adult?

(1) legs
(2) lungs
(3) gills
(4) tail
(5) mouth

22. The tadpole has gills, a caudal fin and a two-chambered heart. Which of the following kinds of animals is it *most* like?

(1) reptile   (2) mammal
(3) amphibian (4) insect
(5) fish

GO ON TO THE NEXT PAGE.

Items 23–28 refer to the following passage and diagram.

The human brain consists of three segments: the cerebrum, the cerebellum, and the brain stem.

The cerebrum is the largest area of the brain. It consists of two halves, or hemispheres. The hemispheres are connected by fibers and nerves. Because the fibers and nerves that connect the hemispheres cross, the left half of the cerebrum controls the right side of the body and vice versa. However, each hemisphere seems to specialize in certain functions. For example, the left hemisphere is largely responsible for language and logical thinking. The right hemisphere is largely responsible for artistic and perceptual expression.

All of the activities of the cerebellum are involuntary and work below the level of consciousness. The cerebellum helps the cerebrum control muscular activity, coordination and muscle tone. It also coordinates information from the eyes, inner ears and muscles to maintain balance.

The brain stem is made up of the pons, which receives information from the facial nerves. The lower part of the brainstem is called the medulla oblongata; it controls the activities of the inner organs.

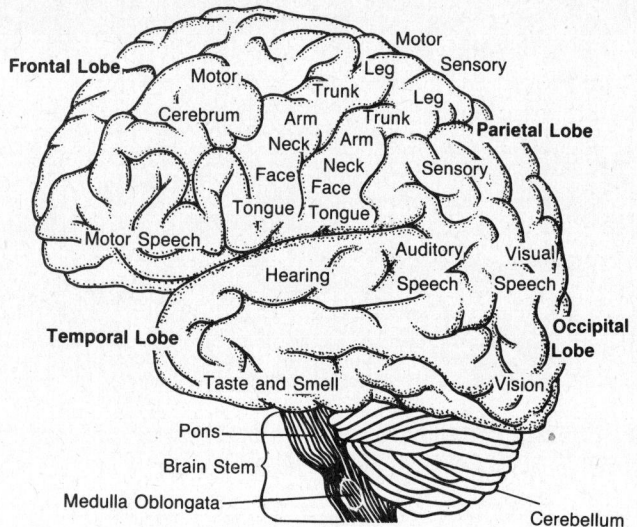

23. Which part of the brain controls respiration?

    (1) pons
    (2) cerebellum
    (3) medulla oblongata
    (4) cerebrum
    (5) frontal lobe

24. According to the diagram, which part of the cerebrum controls hearing?

    (1) temporal lobe
    (2) frontal lobe
    (3) parietal lobe
    (4) occipital lobe
    (5) cerebellum

25. A person's musical ability is *largely* controlled by which hemisphere of the cerebrum?

    (1) left
    (2) right
    (3) temporal
    (4) occipital
    (5) parietal

26. Which of the following functions is *most* likely to be affected by a sharp blow to the back of the head?

    (1) taste
    (2) vision
    (3) heart rate
    (4) motor speech
    (5) leg movement

27. If a person loses control of his or her balance, what part of the brain is probably affected?

    (1) pons
    (2) medulla oblongata
    (3) cerebellum
    (4) cerebrum
    (5) temporal lobe

28. When you view an object through your left eye, which part of your brain is receiving the message?

    (1) right occipital lobe
    (2) left temporal lobe
    (3) right temporal lobe
    (4) cerebellum
    (5) medulla oblongata

GO ON TO THE NEXT PAGE.

Items 29–32 refer to the following passage and graph.

The body has a natural way of preventing disease called the immune response. The immune response is largely controlled by two kinds of lymphocytes called B cells and T cells. The B cells produce and release antibodies for specific diseases. Antibodies are chemicals that attack the toxins, or poisons, released by disease-causing microorganisms. Different kinds of B cells produce different kinds of antibodies. The T cells help the B cells multiply and tell them when to stop producing the antibody.

When disease-causing organisms such as bacteria or viruses infect the body, B cells begin producing antibodies to the specific toxins or antigens released by the organisms. T cells then help the B cells divide, forming a cluster of cells. Each cell in the cluster is capable of producing the antibody. Soon a large quantity of the antibody is secreted into the bloodstream to help fight off the disease. This is called the primary immune response (see graph). When the disease organisms are all destroyed, T cells cause the B cells to stop producing the antibody. The B cells then become inactive. However, if the same disease organism enters the body again, the B cells are ready to secrete the antibody almost immediately, preventing the disease from occurring a second time. This is called the secondary immune response.

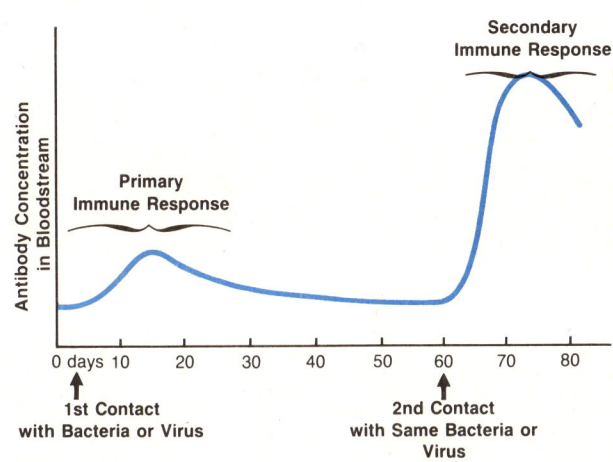

29. The secondary immune response is much more rapid than the primary response because there are

    (1) more T cells in the body
    (2) fewer disease organisms in the body
    (3) B cells ready to produce antibodies
    (4) more disease organisms in the body
    (5) large quantities of antibody already in the bloodstream

30. If a virus was able to prevent B cells from multiplying, what would be the *most* likely effect?

    (1) T cells would begin to multiply.
    (2) Enough antibody could not be produced to fight the disease.
    (3) Too much antibody would be produced.
    (4) The virus would die.
    (5) The secondary immune response would take effect.

31. A person contracts a disease the first time the disease enters the body because there are no

    (1) B cells in the body
    (2) T cells in the body
    (3) lymphocytes in the body
    (4) immune response defenses
    (5) antibodies to fight the disease

32. Which of the following *most* likely explains why a person does not become immune to colds?

    (1) Many different viruses cause colds.
    (2) Colds are caused by the same viruses.
    (3) All colds can be attacked by the same kinds of antibodies.
    (4) T cells are not producing antibodies.
    (5) There are too many B cells.

GO ON TO THE NEXT PAGE.

Items 33–36 refer to the following passage and charts.

There are three different genes that determine a person's blood type. However, an individual only inherits one gene from each parent, so several different gene combinations are possible. The result is that four blood types are found in humans. The system most commonly used to identify blood types is called the A-B-O system. The chart at the right shows the genotype for each blood type. For example, a person with type A blood has either two genes for type A or one type A gene and one type O gene. Type O genes are always recessive, while type A and B genes are dominant. A person who receives a type A gene and a type B gene from his or her parents has type AB blood.

Type	Genotype	Can Get Blood From	Can Give Blood To
A	AA, AO	O, A	A, AB
B	BB, BO	O, B	B, AB
AB	AB	A, B, AB, O	AB
O	OO	O	A, B, AB, O

Since the blood type a person has depends on the two genes inherited from his or her parents, different populations show different frequencies of blood types. The chart below shows the percentage of different blood types among different groups of people. For example, only about 2.2% of Polynesians have type B blood while about 35% of Chinese people have type B blood.

Blood Type	A	B	AB	O
U.S.A.—white	41.0%	10.0%	4.0%	45.0%
U.S.A.—black	26.0%	21.0%	3.7%	49.3%
Swedish	46.7%	10.3%	5.1%	37.9%
Japanese	38.4%	21.8%	8.6%	31.2%
Polynesian	60.8%	2.2%	0.5%	36.5%
Chinese	25.0%	35.0%	10.0%	30.0%
North American Indian	7.7%	1.0%	0.0%	91.3%

33. People with type O blood are sometimes called universal donors. A universal donor is a person who

    (1) can receive blood from anyone
    (2) cannot donate blood to anyone
    (3) has a genotype of OO
    (4) can donate blood to anyone
    (5) can only receive type O blood

34. Which blood type can receive blood of any type?

    (1) type A
    (2) type B
    (3) type AB
    (4) type O
    (5) type AA

35. According to the chart, which group of people has the *highest* percentage of type O blood?

    (1) Swedish
    (2) Japanese
    (3) Polynesian
    (4) Chinese
    (5) North American Indians

36. According to the chart, which blood type is the *least* common?

    (1) type A
    (2) type B
    (3) type AO
    (4) type AB
    (5) type O

GO ON TO THE NEXT PAGE.

Items 37–40 refer to the following map and definitions.

Large regions of the earth with given climates and specific kinds of plants and animals are called biomes. The locations and descriptions of six terrestrial biomes are shown in the map below.

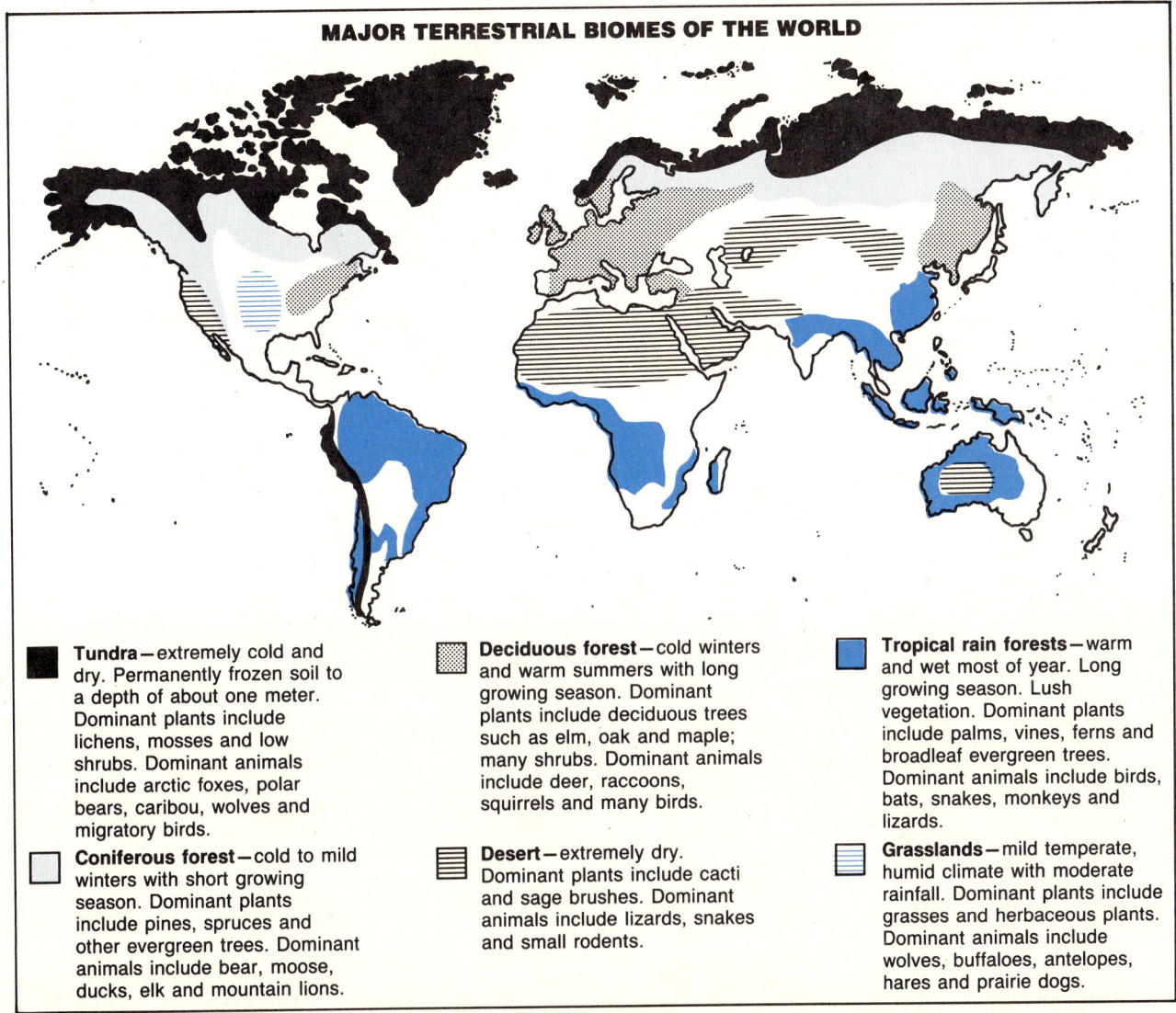

37. According to the map, which kind of biome occurs furthest north?

    (1) grassland
    (2) deciduous forest
    (3) tundra
    (4) coniferous forest
    (5) tropical rain forest

38. What kind of biome makes up *most* of the midwestern United States?

    (1) grassland
    (2) desert
    (3) deciduous forest
    (4) coniferous forest
    (5) tundra

39. Which of the following is characteristic of a tropical rain forest biome?

    (1) cold climate
    (2) abundance of grasses
    (3) short growing season
    (4) warm summers
    (5) large amount of rainfall

40. Which of the following kinds of plants would you *not* expect to find in a tropical rain forest?

    (1) palm trees
    (2) vines
    (3) cacti
    (4) tree ferns
    (5) broadleaf evergreen trees

Check your answers to the Review on pages 153–155.

OVERVIEW
Earth Science

Volcano island of Cracatoa in Indonesia after the 1883 eruption.

era
the largest division of geological time

Earth science is the branch of science that studies the earth and the space around the earth. Knowledge of earth science can be useful in many ways, for it helps us understand and use the environment in which we live.

In the first lesson of this section you will learn about the structure of the earth. You will also learn how the earth has changed over time. Major changes have signaled the beginning or end of an **era**, which is the largest division of geological time.

In the second lesson you will learn about the air that surrounds the earth. The envelope of gases that surrounds the earth is called the **atmosphere.** The atmosphere, which consists of four main layers, extends from the surface of the earth into outer space. The layer of atmosphere most familiar to us is the one in which we live, the troposphere.

In this lesson you will also learn about the water that covers approximately 70 percent of the earth's surface. The earth's water supply is constantly being renewed by a series of steps called the **water cycle.** As you learn about the water cycle, you will come to understand how moisture from the earth forms clouds and how this moisture eventually falls back to earth in the form of rain, snow, sleet or hail.

Air and water create changing weather patterns. You will discover how some of our weather is due to the movement of **air masses,** large areas of air that have uniform temperature and humidity.

In the third lesson you will learn about the earth's resources. You will discover that one of the world's most important resources is petroleum. Petroleum provides fuel for energy, and also provides the raw materials needed for many useful chemicals. Using so much petroleum, however, has presented a problem, because petroleum is a **nonrenewable resource**—that is, it cannot be replaced once it is used up. Many earth scientists fear that the world may run out of petroleum before other resources can be developed to take its place.

Another important resource is our food supply. As you study a **food chain,** you will learn how living things depend upon one another for food. You will also learn how the food supply for certain animals can be damaged by a particular form of pollution known as acid rain.

In the last lesson you will learn how the earth is constantly changing. Some of these changes, such as earthquakes and volcanoes, are dramatic and dangerous. Others, such as **erosion** and **mass wasting,** can occur so slowly that it is not until many years later that we see their effects.

atmosphere
the envelope of gases that surrounds the earth

water cycle
continuous movement of water from water sources to the air, then back to the water sources again

air mass
large body of air that has the same temperature and moisture throughout

nonrenewable resource
resource that cannot be replaced once it is used up

food chain
illustration of how groups of organisms are dependent upon one another for food

erosion
the wearing away and moving of rock materials by natural agents

mass wasting
the downhill movement of sediments caused by gravity

The Planet Earth

The surface of the earth covers about 510,000,000 square kilometers. Yet this area represents only a small part of the whole earth; most the earth lies buried beneath our feet.

For many years, scientists have been gathering information about the earth's interior. Intense heat and high pressure make direct exploration of this region impossible, so most of what is known about the interior structure of the earth has been learned by studying the movement of seismic waves that are produced by earthquakes. Based on the study of seismic waves, scientists have concluded that the earth is made up of four different layers: the crust, the mantle, the outer core and the inner core.

The **crust** is the part of the earth familiar to us, for it includes the surface of the earth. The crust is about 8 kilometers thick under the oceans and about 32 kilometers thick under the continents. Under mountains the crust may extend to a thickness of 70 kilometers. The crust is made up of different kinds of rock. Beneath the crust lies the **mantle.** The mantle is composed of rock that contains primarily oxygen, iron and silicon. The temperatures in this region—870°C to 2,200°C—cause some of the solid rock to flow like a liquid. Below the mantle is the **outer core,** which begins 2,900 kilometers beneath the earth's surface and extends down another 2,250 kilometers. The outer core is composed of molten iron and nickel. Temperatures in the outer core range from 2,200°C to 5,000°C. At the center of the earth is the **inner core.** This region, which has a temperature of about 5,000°C, is solid iron and nickel. Although iron and nickel will usually melt at this temperature, great pressure in the inner core pushes the particles together so tightly that they remain solid. The presence of a dense inner core of iron may help explain why a magnetic field surrounds the earth.

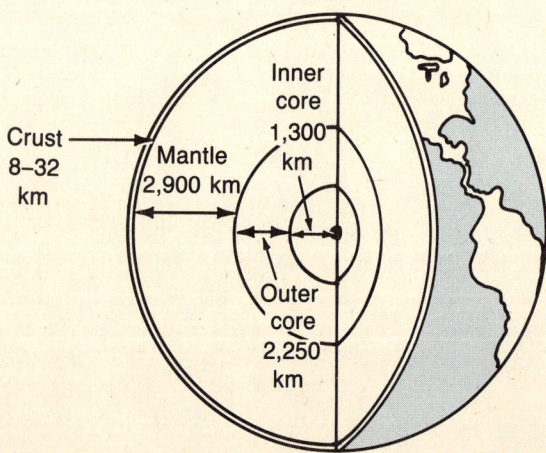

Strategies for READING

Identify an Implication

This skill identifies assumptions, facts or statements that are taken for granted (not proved), and that the author takes for granted.

You can increase your ability to spot **implications** if you think about these three strategies as you read:

Think about consequences. A consequence is an effect or result. The passage on the previous page states that the interior of the earth contains "intense heat and pressure." Since high heat and pressure are destructive to human beings, it is implied that humans attempting to probe the earth's interior would be destroyed.

Be aware of inference. To infer is to make a reasonable assumption based on the information given. Since the passage states that direct exploration of the inner earth is impossible, it is logical to assume that no one has ever traveled beneath the crust of the earth.

Look for generalizations. If a statement is true in general, it must be true for specific items as well. If the thickness of the earth's crust is approximately 32 kilometers under the continents, then it must be 32 kilometers under Australia.

Examples

DIRECTIONS: Use the information on this and the preceding page to choose the one best answer for each item below.

1. What is the *most* likely measure of the earth's crust under the Indian Ocean?

    (1) about 1 km
    (2) about 8 km
    (3) about 30 km
    (4) about 55 km
    (5) about 70 km

Answer: (2) The implication is that since the thickness of the earth's crust is "about 8 kilometers" under the oceans (generalization), it is about 8 kilometers under the Indian Ocean (specific).

2. Which of the following *best* describes indirect evidence?

    (1) was obtained by accident
    (2) takes a long time to collect
    (3) is difficult to collect
    (4) describes earthquakes
    (5) was obtained without seeing or touching the object being studied

Answer: (5) The inference is that scientists gained information about parts of the earth that could not be explored directly, so the evidence was obtained without seeing or touching the object being studied.

READING: COMPREHENSION

Practice

> **HINT**: Remember to think about consequences, inferences and generalizations as you read. This will help you spot implications, facts or statements that are taken for granted.

DIRECTIONS: Choose the <u>one</u> best answer for each item below.

Items 1–4 refer to the following passage.

A **fossil** is the evidence or remains of a living thing. Most fossils form when plants or animals die and are buried in sediments that harden into rock.

The chances of an organism leaving a fossil are actually slight. The soft parts of a dead organism tend to decay or be eaten by animals before a fossil can form. Fossils that do form are often incomplete. Most likely to be preserved as fossils are plants or animals that lived in or near water. There, sediments of sand and mud provided quick burial for these organisms.

Some fossils show only the mark or evidence of a living thing. Called **trace fossils**, these fossils are formed from footprints, tracks and burrows. Much of what is known about the dinosaur has come from footprints found in rock.

Fossils can indicate changes in the earth's surface and climate. For example, fossils of coral found in Antarctica are an indication that the climate of this region was once considerably warmer.

Fossils can also help date layers of rock. If a particular type of organism lived on the earth for only a brief period of time, the rock containing the fossil of the organism must be from approximately the same time period.

1. In a mountainous region of Canada, fossils of fish are found. Which is the *most* likely explanation?

    (1) The region was once much colder.
    (2) The region was once much warmer.
    (3) The region was once underwater.
    (4) Fish that lived in mountain streams formed fossils.
    (5) Dead fish were taken up the mountain.

2. Which of the following would be *most* similar to a trace fossil?

    (1) human footprints made in concrete before it hardened
    (2) a fly trapped in concrete before it hardened
    (3) tire tracks in the sand
    (4) footprints of a deer in the snow
    (5) tire tracks made in concrete before it hardened

GO ON TO THE NEXT PAGE.

3. Sedimentary rock forms in layers. In a certain region, one layer of sedimentary rock contains the fossil of a bear and the layer beneath it contains the fossil of an alligator. Which of the following could be said about this region?

   (1) The climate has remained essentially the same.
   (2) The climate was once warm, then became cold.
   (3) The climate was once cold, then became warm.
   (4) The terrain was once mountainous, then became flat.
   (5) The terrain was once flat, then became mountainous.

4. The passage implies that fossils are often incomplete because

   (1) the fossils are trace fossils
   (2) the climate of the region changed drastically
   (3) the organisms were trapped in sand and mud
   (4) the organisms lived for only a short period of time
   (5) parts of the organisms decayed or were eaten by animals

Items 5–8 refer to the following time-line.

Scientists have designed a geological time-line to record the history of the earth.

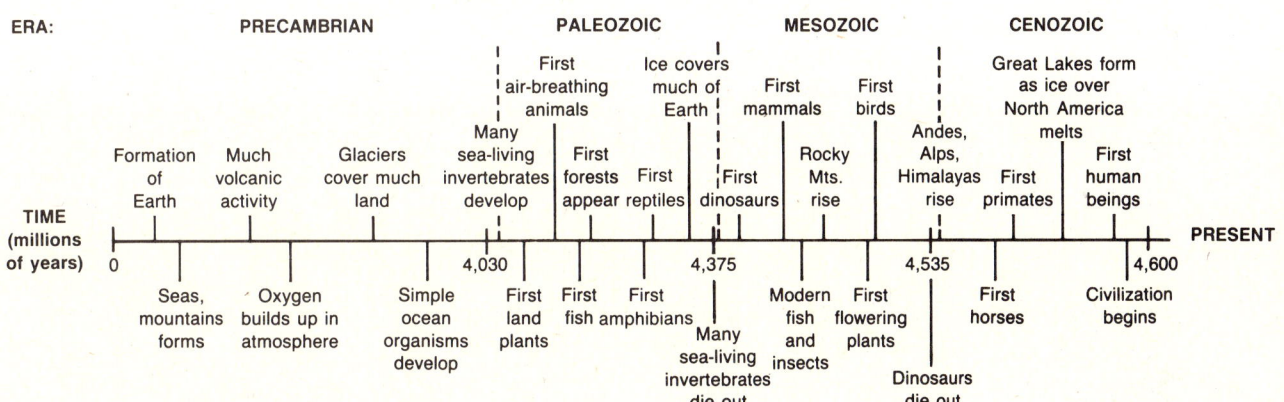

5. According to the time-line, where did the first living things originate?

   (1) in a volcano
   (2) in the atmosphere
   (3) in the ocean
   (4) in a swampland
   (5) in the mountains

6. For approximately how many years did dinosaurs inhabit the earth?

   (1) 160 years
   (2) 4,375 years
   (3) 160 million years
   (4) 160 billion years
   (5) 4,535 billion years

7. The Paleozoic Era was characterized by which of the following?

   (1) the formation of glaciers
   (2) the appearance and extinction of many sea invertebrates
   (3) the extinction of reptiles
   (4) the appearance of the first flowering plants
   (5) the appearance of the first horses

8. Human civilization has been in existence for how long?

   (1) since the beginning of the earth
   (2) since the end of the Paleozoic Era
   (3) since the middle of the Mesozoic Era
   (4) since the beginning of the Cenozoic Era
   (5) since the most recent part of the Cenozoic Era

Before you take the GED Mini-Test, check your answers on pages 85–86.

GED Mini-Test 9

TIP If you skip a question, keep track of where you are on your answer sheet. Be sure to correctly place your answer to each test item next to its corresponding test item number.

DIRECTIONS: Choose the one best answer for each item below.

Items 1–4 refer to the following passage.

The earth's crust contains three main types of rock: sedimentary, igneous and metamorphic. Sedimentary rock is formed when sediments of sand, mud, pieces of rock or remains of dead organisms harden. Igneous rock forms when hot liquid rock, called magma, hardens into crystals. Metamorphic rock results when rocks are subjected to extreme heat and pressure. Sedimentary rocks are formed near the surface of the earth, usually in water. Igneous and metamorphic rocks are formed in the lower part of the earth's crust or in the upper part of the mantle.

As magma, or molten rock, rises it tends to cool. It can cool above or below the surface of the earth. Large mineral crystals form when magma cools slowly. Heat is lost very slowly beneath the earth's surface. Magma can break through the earth in a volcanic eruption. It is then called lava. As lava comes in contact with the air, cooling can be very rapid. This sudden cooling leads to the formation of barely visible crystals.

1. According to the information provided, which of the following could explain why igneous and metamorphic rocks are formed deep beneath the earth's surface?

   (1) Temperatures in the earth's interior are high.
   (2) Pressure in the earth's interior is low.
   (3) The rocks are protected from wind and weather.
   (4) Chemical reactions can occur beneath the earth's crust.
   (5) Magma hardens more quickly beneath the earth's crust.

2. Suppose an igneous rock reaches the earth's surface and is broken down by weathering and erosion. Which of the following could happen to the pieces of rock?

   (1) They could melt and become magma.
   (2) They could become sediments and harden into sedimentary rock.
   (3) They could be pushed together to form a metamorphic rock.
   (4) They could become hot and form another igneous rock.
   (5) They could be carried into the ocean and form metamorphic rock.

3. Which *best* explains the presence of large mineral crystals?

   (1) the force of a volcano
   (2) the slow cooling of molten rock in the earth's crust
   (3) the effect of air on the temperature
   (4) the rapid cooling of lava
   (5) the pressure on magma

4. Which *best* explains why small mineral crystals are likely to form?

   (1) contrast in temperature between lava and air
   (2) lava is molten rock above the surface of the earth
   (3) cooling is slower beneath the earth's surface
   (4) volcanic eruptions break rocks into small pieces
   (5) molten rock rises then cools

GO ON TO THE NEXT PAGE.

Items 5–6 refer to the following passage.

In order to be classified as a mineral, a substance must have five basic properties: It must occur naturally in the earth; it must be a solid; it must be inorganic; it must have definite chemical composition; and its atoms must be arranged in a definite pattern called a crystal. Rock deposits that contain valuable minerals are called ores. In order to extract the mineral, the ores must first be mined, then subjected to a process called smelting. During smelting, the ore is heated in such a way that the mineral is separated from the impurities.

5. Which of the following could *not* be a mineral?

   (1) a crystalline salt
   (2) a substance formed in the mantle
   (3) a substance consisting of 18% oxygen and 82% silicon
   (4) a solid that is shiny
   (5) a rock formed from plant remains

6. Which of the following would be involved in removing iron from its ore?

   (1) adding impurities
   (2) polishing
   (3) freezing
   (4) high temperatures
   (5) strong odors

Check your answers to the GED Mini-Test on page 86.

Answers and Explanations

Practice   pp. 82–83

1. **Answer:** (3)   Since fossils are most likely to be formed by organisms in or near the water, the best choice is (3). Choices (1) and (2) are irrelevant. Choice (4) is remotely possible, but since mountain streams are usually shallow and rocky, it is unlikely that enough soft sediment would be available to quickly bury the organism. Choice (5) is far-fetched—surely the fish would decay or be eaten before they made it up the mountain.

2. **Answer:** (1)   The best choice is (1) because it describes evidence of living activity (human footprint) in a medium that will harden and become permanent (concrete). Choices (2) and (5) are incorrect since a fly is a whole organism, not a trace, and a tire is not a living thing. Choices (3) and (4) are incorrect because sand and snow are not permanent mediums.

3. **Answer:** (2)   The more recent layer of rock would be the top layer. Since an alligator is a tropical creature (bottom layer) and the bear is a cold-climate animal (top layer), the climate must have changed from warm to cold. Choices (1) and (3) contradict the correct answer, and choices (4) and (5) are irrelevant.

4. **Answer:** (5)   Since decay and consumption by animals prevent organisms from forming fossils, it is logical to assume that these same processes would prevent the formation of complete fossils. Choices (1) through (4) take information from the passage out of context—a common practice in multiple-choice tests and something you should watch out for!

5. **Answer:** (3)   Since the first organisms mentioned on the time-line are simple ocean organisms, the first living things must have originated in the ocean. None of the locations mentioned in choices (1), (2), (4) or (5) are connected with a particular organism.

6. **Answer:** (3)   This question requires you to read the time-line carefully. The time-line is measured in *millions* of years. Subtracting 4,375 from 4,535 gives 160, which on this time-line represents 160 million years. If you answered choice (1) or (4), you misread the scale. If you answered choice (2) or (5), you forgot to subtract.

7. **Answer:** (2) The time-line indicates that many sea invertebrates appeared at the beginning of the Palezoic Era and had died out by the end of the era. Choices (1), (4) and (5) are not relevant to this era. Reptiles, choice (3), appeared in this era, but did not die out.

8. **Answer:** (5) This time-line shows human beings appearing at the very last part of the Cenozoic Era, indicating that human civilization is relatively recent in the earth's history. All other answer choices indicate times that are too early.

GED Mini-Test *pp. 84–85*

1. **Answer:** (1) According to the passage, igneous rock is formed from hot liquid rock, and metamorphic rock is formed in the presence of high temperature and pressure. Since both types of rock require high temperatures, which are found in the earth's interior, the correct choice is (1). Choice (2) contradicts the information presented about metamorphic rock, and is also an untrue statement about the earth's interior. Choices (3) and (4) are irrelevant. Choice (5) is vague and would pertain only to igneous rock.

2. **Answer:** (2) Since only sedimentary rock is formed on or near the earth's surface, only choice (2) is correct. Choices (1), (3) and (4) could be correct if the rock fragments were buried deep in the earth. Choice (5) is incorrect because it is sedimentary, not metamorphic, rock that is formed in the ocean.

3. **Answer:** (2) Choices (1), (3) and (4) are factors that do not contribute to the slow cooling of molten rock, or magma, which is necessary for the formation of large mineral crystals. Choice (5) is a factor not mentioned in the passage.

4. **Answer:** (1) Choices (2) and (3) are facts, but do not offer an answer to the question. Choice (4) has no relationship to the question. Choice (5) offers a reason why magma tends to cool off.

5. **Answer:** (5) Since plant remains are organic, a rock formed from plant remains could not be classified as inorganic. Choices (1) through (4) could all be minerals since they include the properties crystal, choice (1), found in the earth, choice (2), definite chemical composition, choice (3), and solid state, choice (4).

6. **Answer:** (4) Since smelting is a process that involves heating, one can assume that high temperatures are involved. Choice (1) is incorrect because smelting removes impurities. Choice (2), polishing, might be involved in eventually preparing iron for use, but it would not be involved in separating the iron from its ore. Choice (3) is incorrect because the process requires heating, not cooling. Choice (5) is irrelevant, but it may appear correct to some people who think that smelting must have something to do with the sense of smell.

Air and Water

Two things that are essential to human life are air and water. In this lesson you will read about the earth's atmosphere. Later in this lesson you will learn about the earth's fresh water and oceans.

The earth's atmosphere provides us with a safe, comfortable environment. It gives us moisture and oxygen, a comfortable temperature and protection from the sun's ultraviolet rays.

The atmosphere contains gases that are necessary for the survival of all living things. These gases include nitrogen, oxygen, carbon dioxide, water vapor and argon. Present in very small amounts are neon, helium, krypton and xenon.

The atmosphere is divided into four main layers. The lowest layer is called the **troposphere.** This is the layer in which we live. The troposphere extends to a height of about 16 kilometers, or approximately 10 miles. The temperature and density of the troposphere change as distance from the surface of the earth changes. As altitude increases, the air becomes colder and less dense, or "thinner." For example, at an altitude of 5.5 kilometers, or about 3.5 miles, there is only half as much oxygen available as there is at the earth's surface.

Above the troposphere is the **stratosphere.** The stratosphere extends to a height of about 50 kilometers, or about 31 miles. A special form of oxygen called **ozone** is present in the stratosphere. It is ozone that shields the earth from the sun's harmful ultraviolet rays.

Above the stratosphere is the **mesosphere.** The mesosphere extends to about 80 kilometers, or 50 miles, above the earth's surface. The temperature in the mesosphere drops to nearly −100°C.

The uppermost region of the atmosphere is the **thermosphere.** The thermosphere has no well-defined upper limit. It is the hottest layer of the atmosphere, with temperatures as high as 2000°C.

NASA view of Earth from the Apollo 8 spacecraft.

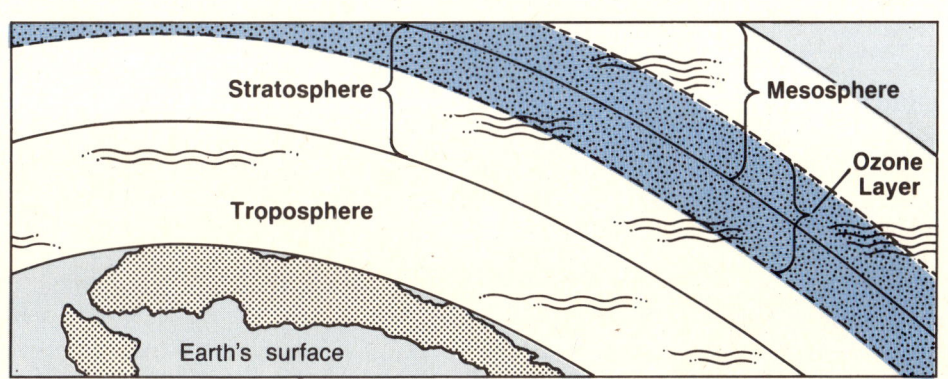

INTRODUCTION 87

Strategies for READING

Use Given Ideas in Another Context

This skill involves taking information from a passage and using it in another situation.

One of the most practical subjects you will study is science. Science explains the physical world around you and within you.

When you read science, it is important to recognize how the given information can be used in another **context.** Often a test question will ask you to make a practical application of what you have read. This application could be to a situation in daily life, or one relating to a familiar aspect of science. For example, on the previous page you read about the earth's atmosphere. A test question based on this passage might require you to recognize that a spacecraft traveling through all four layers of the atmosphere would have to be able to withstand dramatic changes in temperature.

You can increase your ability to apply information in other contexts by asking yourself these questions as you read:

1. What is being described or explained?
2. What types of situations would relate to this topic?
3. How might the given information be useful in those situations?

Examples

DIRECTIONS: Use the information on this and the preceding page to choose the one best answer for each item below.

1. An astronaut traveling beyond the earth's atmosphere would need protection from which of the following?

    (1) ozone
    (2) moisture
    (3) the sun's ultraviolet rays
    (4) poisonous gases
    (5) dense clouds

    Answer: (3)   Choices (1) and (2) are incorrect because moisture and ozone provide comfort and protection *within* the atmosphere. Choices (4) and (5) are not mentioned in the passage.

2. Which of the following would a person climbing a mountain *most* likely expect to encounter?

    (1) hail
    (2) dry air
    (3) heavy rain
    (4) low temperatures
    (5) high temperatures

    Answer: (4)   Air becomes colder, not warmer, as altitude increases. Choices (1), (2) and (3) do not relate to the information given.

Practice

HINT: Underline key words or ideas in each paragraph as you read. Use your underlining to help you locate information quickly as you answer each question.

DIRECTIONS: Choose the one best answer for each item below.

Items 1–4 refer to the following article.

Changes in weather are caused by movements of air called **air masses.** Air masses cover thousands of square kilometers and have the same temperature and humidity throughout.

Air masses are classified according to where they form. There are four major types of air masses that affect the United States: maritime tropical, maritime polar, continental tropical, and continental polar. They are called maritime if they come from the sea and continental if they originate over land.

A **maritime tropical** air mass forms over the ocean near the equator. Its air is warm and moist. In the summer it brings hot, humid weather, but in the winter it may come in contact with a cold air mass and cause rain or snow to fall.

A **maritime polar** air mass forms over the Pacific Ocean in both winter and summer, and over the North Atlantic Ocean in summer. It contains cool, moist air. During the summer, this air mass brings fog to the western coastal states and cool weather to the eastern states. In the winter, heavy snow and very cold temperatures are produced by this air mass.

A **continental tropical** air mass forms over land in Mexico during the summer. It brings hot, dry air to the southwestern states.

A **continental polar** air mass forms over land in northern Canada. It contains cold, dry air. It brings very cold weather to the United States in winter.

1. Which of the following statements describes air masses that form over land?

   (1) They contain cold, moist air.
   (2) They contain warm, moist air.
   (3) They contain dry air.
   (4) They contain moist air
   (5) They contain moist air in winter and dry air in summer.

2. A blizzard in the Northern Pacific States could be caused by which of the following air masses?

   (1) maritime polar
   (2) continental tropical
   (3) continental polar
   (4) maritime polar or continental polar
   (5) continental polar or continental tropical

GO ON TO THE NEXT PAGE.

PRACTICE 89

3. A person flying cross-country leaves cool summer weather in New York and arrives to find fog in Los Angeles. Which of the following statements could describe the air masses affecting these areas?

    (1) Maritime tropical air masses affect both areas.
    (2) Maritime polar air masses affect both areas.
    (3) Continental polar air masses affect both areas.
    (4) A maritime polar air mass affects New York while a continental tropical air mass affects Los Angeles.
    (5) A continental polar air mass affects New York while a maritime polar air mass affects Los Angeles.

Items 5–8 refer to the diagram.

5. Between which two factors does the diagram illustrate a relationship?

    (1) distance from the Equator and angle of the sun's rays
    (2) distance from the Equator and length of day and night
    (3) angle of the sun's rays and weather patterns
    (4) angle of the sun's rays and length of day and night
    (5) distance from the Equator and average rainfall

7. According to the diagram, which of the following statements is true?

    (1) The sun's rays strike all parts of the earth evenly.
    (2) The sun's rays are most direct at the North Pole.
    (3) The sun's rays are most direct at the South Pole.
    (4) The sun's rays are most direct in the Northern Hemisphere.
    (5) The sun's rays are most direct at the Equator.

4. Which of the following is a *false* statement concerning air masses?

    (1) An air mass is called maritime because it originates over water.
    (2) An air mass is called continental because it originates over land.
    (3) Air masses called tropical originate over tropical seas.
    (4) An air mass can cause a change in humidity.
    (5) Air masses cover a very large area.

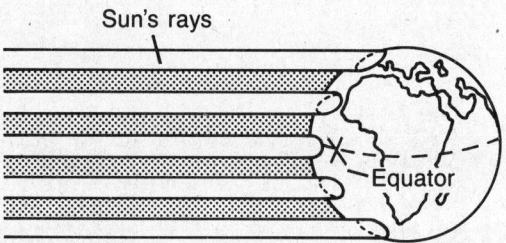

6. Climates near the Equator are warmer than climates near the poles. Based on the diagram, which of the following statements could explain this phenomenon?

    (1) Days are longer near the Equator than near the poles.
    (2) The sun's rays are more concentrated near the Equator than near the poles.
    (3) Weather patterns produce more sunny days near the Equator than near the poles.
    (4) There are fewer oceans near the Equator than near the poles.
    (5) There is less wind near the Equator than near the poles.

8. What should a person traveling from the middle of the Northern Hemisphere to the Equator be prepared for?

    (1) shorter days
    (2) shorter nights
    (3) cooler temperatures
    (4) a greater possibility of rainy weather
    (5) a greater possibility of sunburn

GED Mini-Test 10

TIP Read all the choices on the GED test *before* making your selection. Rule out the obviously incorrect choices, probably three of the five choices. You will be increasing your chances of selecting the correct answer.

DIRECTIONS: Choose the one best answer for each item below.

Items 1–4 refer to the following passage.

The earth's supply of fresh water is constantly being renewed by the water cycle. The water cycle consists of three steps.

The first step is evaporation. Heat from the sun causes large amounts of water on the surface of the earth to change into water vapor. Much of this water comes from the oceans; the rest comes from the soil, vegetation and fresh water sources. The water vapor is carried by winds over land and oceans.

The next step in the cycle is condensation. Condensation is the process by which vapor turns back into liquids. This happens when warm air near the surface of the earth rises and is cooled. Once the air cools, it can no longer hold as much water vapor. The excess water vapor condenses into droplets that form clouds.

The final step in the water cycle is precipitation. When the droplets of water become too heavy and too numerous to float in air, water is released in the form of rain or snow. Much of this water finds its way back into oceans, rivers, lakes and soil. Now this water is warmed by the sun. As some of it evaporates, the water cycle begins all over again.

1. Which of the following statements *best* explains why clouds form?

   (1) Water vapor in the air condenses.
   (2) Droplets of water in the air become heavy.
   (3) Droplets of water in the air evaporate.
   (4) Heat from the sun reacts with gases in the atmosphere.
   (5) Warm air above the earth reacts with cool air on the surface of the earth.

2. Rain or snow is produced when which of the following sequences of events takes place?

   (1) Water vapor condenses, then evaporates.
   (2) Air becomes cold, then becomes warm.
   (3) Water evaporates, then condenses.
   (4) Water condenses, then is carried by wind.
   (5) Air rises, then sinks.

3. What is the *best* explanation of why the water cycle is constantly being renewed?

   (1) Clouds release precipitation.
   (2) Oceans and rivers carry water.
   (3) Evaporation and condensation cause precipitation.
   (4) Precipitation causes condensation.
   (5) Evaporation causes condensation.

4. What would *most* likely occur if the water cycle was not constantly renewed?

   (1) The sun would change how plant life breathes.
   (2) The sun would change how animal life breathes.
   (3) Animal life would thrive.
   (4) Plant and animal life would die.
   (5) Plant and animal life would be carried away by oceans.

GO ON TO THE NEXT PAGE.

Items 5–6 refer to the following passage.

Approximately 70% of the earth's surface is covered by oceans. Ocean water makes up 97% of all the water found on earth. Ocean water is a mixture of gases and solids dissolved in pure water. The most abundant salt in ocean water is sodium chloride. Sodium chloride is common table salt.

Scientists use the term salinity to describe the saltiness of ocean water. Salinity is a measure of the amount of salt per 1,000 kilograms of water. Most ocean water has a salinity of between 33 and 37, meaning that between 33 and 37 grams of salt will be found in 1,000 kilograms of water. The salinity tends to be highest in places where there is a great deal of evaporation or little rainfall.

The sun is the major source of heat for the ocean. As a result, temperatures are highest at the surface of the ocean. Waves and currents transfer the heat downward to a depth of about 300 meters. Below 300 meters, the temperature of ocean water drops rapidly. By the time a depth of 1,500 meters has been reached, the temperature of the water is just a few degrees above freezing.

5. Most living things cannot use water that contains salt. According to the information provided, which of the following statements must be true?

   (1) About 70% of the water on earth can be used to meet the needs of living things.
   (2) Most of the water on earth cannot be used to meet the needs of living things.
   (3) Pollution has severely reduced the amount of water that can be used by living things.
   (4) The amount of water that can be used by living things is steadily increasing.
   (5) Lack of rainfall has decreased the amount of water available to living things.

6. Which of the following situations is *most* like the heating of ocean water?

   (1) A frozen dinner placed in a microwave oven is heated evenly throughout.
   (2) People sitting near a fireplace feel warm, while people sitting far from the fireplace feel chilly.
   (3) Water heated on a stove begins to boil when it reaches a certain temperature.
   (4) An electrical wire becomes hot when too much current flows through it.
   (5) A thermos bottle keeps liquids hot for many hours on a cold day.

Check your answers to the GED Mini-Test on page 93.

Answers and Explanations

Practice pp. 89–90

1. **Answer:** (3) Choices (1), (2), (4) and (5) are incorrect because continental air masses contain dry air only.

2. **Answer:** (1) Choices (2), (3), (4) and (5) are incorrect because continental air masses do not cause precipitation. The only other type of air mass that might have caused the snowstorm would be a maritime tropical in winter.

3. **Answer:** (2)   Choice (1) is incorrect because in summer maritime tropical air masses bring hot, humid weather. Choices (3) and (4) are incorrect because fog would not be caused by a continental air mass, which contains dry air. Choice (5) might be plausible, but the article makes no mention of the effect of a continental polar air mass in summer—therefore, this is not the best answer.

4. **Answer:** (3)   Since a continental tropical air mass originates over land, all tropical air masses do not originate over tropical seas. Choices (1) and (2) explain why the terms maritime and continental are used. Choices (4) and (5) are accurate descriptions of air masses.

5. **Answer:** (1)   Choices (2), (3), (4) and (5) are incorrect because the length of day and night, weather patterns and rainfall cannot be determined from the diagram.

6. **Answer:** (2)   Choices (1), (4) and (5) are incorrect because they are not relevant to the diagram. Choice (3) might be true, but there is no evidence given in the diagram to suggest that more direct sunlight produces more clear days.

7. **Answer:** (5)   Choices (1) through (4) are incorrect because the diagram clearly shows that except at the Equator, the sun's rays are dissipated over a wider area because of the curve in the earth's surface.

8. **Answer:** (5)   Choices (1) and (2) are incorrect because length of day and night depend upon the season. Choice (3) is incorrect because direct sunlight at the Equator produces warmer temperatures. Choice (4) is irrelevant.

GED Mini-Test   *pp. 91–92*

1. **Answer:** (1)   Choice (2) is incorrect because heavy droplets in clouds result in rain. Choice (3) is incorrect because water on the surface of the earth evaporates to form water vapor in the air. Choice (4) is irrelevant. Choice (5) is incorrect because warm air, not cool air, is near the surface of the earth.

2. **Answer:** (3)   Choices (1) and (2) are incorrect because the steps are in reverse order. Choice (4) is incorrect because it is water vapor that is carried by the wind. Choice (5) is incorrect because once warm air rises, the water vapor condenses.

3. **Answer:** (3)   This is the only choice that includes all three essential steps of the water cycle. Choices (1), (2) and (5) are only parts of this cycle. Choice (4) has the steps in reverse order.

4. **Answer:** (4)   This choice indicates how essential water is to animal and plant life. Choices (1) and (2) are concerned with another cycle. Choice (3) is obviously absurd, and choice (5) is irrelevant.

5. **Answer:** (2)   Choice (1) is incorrect because 70% is a statistic taken out of context—something to beware of on multiple-choice tests! Choices (3), (4) and (5) could be true, but they are not relevant to the information provided.

6. **Answer:** (2)   This is correct because radiant heat from the fire is similar to radiant heat from the sun. The most warmth will be provided to that which is closest to the source of heat. Choice (1) is incorrect because ocean water is not heated uniformly throughout. Choice (3) is incorrect because ocean water is not heated from the bottom up. Choice (4) is not relevant because it deals with electrical energy, not water. Choice (5) is incorrect because the passage does not suggest that ocean water is like an insulator.

11 The Earth's Resources

The earth's resources provide us with the things we need for life, such as food, water and energy. In this lesson you will learn about the earth's resources and how people have a responsibility for using them wisely.

Perhaps some of you remember the gasoline lines of the 1970s. Oil shipments to our country had been drastically cut, and all petroleum products—including gasoline—were in short supply. According to energy planners and earth scientists, an oil shortage could occur again, and, if it did, it could be permanent.

Petroleum, or oil, is what is known as a nonrenewable resource. A **nonrenewable resource** is one that cannot be replaced once it is used up. Earth scientists disagree as to just how much oil is left on earth. Some say that, given our present rate of consumption, the United States will be out of oil by the year 2060. Others feel that there is enough oil to last another 300 years. All agree, however, that at some point the supply of oil will be gone.

Energy planners are trying to find ways to avert a major oil shortage. Most believe that the best solution would be to replace oil with a resource that is renewable. The most abundant renewable resource on earth is radiant energy from the sun. Scientists have developed solar cells that can convert sunlight into electricity. They have also developed ways to heat homes with solar energy. At the present time, however, solar energy cannot be produced economically or on a large enough scale to meet our energy needs.

Another way to avert an oil shortage is to use oil at a much slower rate. Immediately following the oil shortage of the 1970s, the United States passed laws to reduce the maximum driving speed on highways and to lower the thermostats in public buildings in an effort to conserve energy. Citizens were educated about ways to conserve energy, but lifelong habits are hard to break. Many people seem to continue to use old, energy-wasting habits.

A solar energy panel collecting the sun's rays.

INTRODUCTION

Strategies for READING

Distinguish Fact from Opinion

This skill involves determining which statements can be proved.

A **fact** is a statement that can be proved. An **opinion** is a statement that reflects a belief or preference. An opinion cannot be proved. Words and phrases that can help you identify opinions are: believe, feel, seems, agree and could happen.

Much of what you will read in science is fact. These facts have been proven by measurement or by numerous experiments. Sometimes, however, you will come across an opinion. The opinion may be that of the author, or of certain scientists. In the article on the previous page, many opinions are woven together with fact. For example, it is a fact that oil is a nonrenewable resource. It is an opinion that the United States will experience a serious oil shortage at a certain future time.

Examples

DIRECTIONS: Use the information on this and the preceding page to choose the one best answer for each item below.

1. Which of the following statements represents an opinion of some scientists?

    (1) The earth's supply of oil is limited.
    (2) The reduction of driving speeds reduced energy consumption.
    (3) There is enough oil to last 300 years.
    (4) Oil became more plentiful in the 1980s than it had been in the 1970s.
    (5) Solar energy cannot currently meet our demands for energy.

    Answer: (3)   Choice (3) cannot be proved so it is an opinion. Choice (1) is a fact because oil is known to be a nonrenewable resource. Choices (2) and (4) are incorrect because they are facts. Both energy consumption and the availability of oil can be measured. Choice (5) is also a fact.

2. Which of the following statements represents an opinion of the author?

    (1) Americans are wasteful of energy.
    (2) An oil shortage could easily happen again.
    (3) By the year 2060, the United States will be out of oil.
    (4) Solar energy is a renewable resource.
    (5) Information about energy conservation is available in the United States.

    Answer: (1)   Choices (2) and (3) are incorrect because they reflect the opinions of scientists and energy planners, not the author. Choices (4) and (5) are incorrect because they are facts.

READING: ANALYSIS

Practice

HINT As you read, be aware of references to past, present and future. Phrases like previously, at one time and throughout history refer to the past. Words like predict, plan and expect refer to the future. Now and today refer to the present.

DIRECTIONS: Choose the one best answer for each item below.

Items 1–4 refer to the following passage.

The air around the earth is in constant motion. These air movements are called **wind.** Throughout history, people have used energy from the wind to power ships, turn mill wheels and pump water.

Windmills began to appear on American farms around 1860. The energy from these windmills was used to pump water out of the ground for crops and farm animals. In 1890 a windmill was invented that could generate electricity. These wind generators became very popular with American farmers and were widely used.

One problem with wind generators was that they were not always reliable. On calm days they would not work. On stormy days they were often knocked down or blown apart. Because of this, most wind generators were abandoned in the 1940s, when electricity became available from hydroelectric power plants.

The need to find alternatives to fossil fuels, such as coal, oil and natural gas, has sparked new interest in wind energy. In recent years, the use of new materials and designs has made possible the development of several tough, efficient wind generators. These machines can adjust to changing wind conditions and withstand storms.

Energy planners do not expect wind energy to ever meet all of our energy needs. But they do feel that the use of wind energy can help to conserve fossil fuels and reduce air pollution.

1. According to the article, modern wind generators differ from earlier wind generators in that they are

    (1) larger
    (2) less expensive
    (3) able to generate electricity
    (4) more reliable
    (5) used mainly on farms

2. Which of the following aspects of wind energy is *not* discussed in the article?

    (1) how wind generators were used in the 1800s
    (2) how modern wind generators are used
    (3) the reliability of wind generators
    (4) the ability of wind energy to meet energy needs
    (5) the use of wind energy on American farms

GO ON TO THE NEXT PAGE.

3. According to the article, renewed interest in wind energy is a result of which of the following?

(1) a desire to return to a simpler lifestyle
(2) an interest in American history
(3) a need to find alternative energy sources
(4) a desire to better understand wind
(5) an increased need for electricity on American farms

4. Which of the following can we expect to see in the future if the recommendations of energy planners are heeded?

(1) Wind energy will replace fossil fuels as the leading energy resource.
(2) Wind energy will be abandoned.
(3) Wind energy will be the dominant energy resource on the American farm.
(4) Wind energy will reduce the need for electric power.
(5) Wind energy will improve air quality.

Items 5–8 refer to the following passage.

Air pollution is a serious problem facing modern society. One source of air pollution is the burning of coal and oil by factories and power plants. Coal and oil contain sulfur impurities. When these fuels are burned, sulfur is released into the atmosphere. The sulfur reacts with oxygen to form sulfur oxides. Some of these sulfur oxides combine with water in the air to form acids. Eventually these acids fall to the earth as acid rain.

Normal rainwater is neutral or slightly acidic. Acid rain, however, is nearly as acidic as pure lemon juice. When acid rain falls into a lake, much of the plant and animal life dies. Today many lakes look clear and blue because the water is nearly devoid of wildlife. Animals living in the vicinity of such a lake may be dying from starvation.

What can be done about acid rain? Factories and power plants must stop burning sulfur-containing fuels. But fuels with a low sulfur content are often expensive and hard to find.

5. According to the passage, the *primary* cause of acid rain is

(1) lack of oxygen in the air
(2) lack of moisture in the air
(3) high temperatures
(4) burning of certain fuels
(5) discharge of oil into the atmosphere

6. According to the passage, some lakes look clear and blue because they have

(1) an excess of oxygen
(2) an abundance of plants and animals
(3) had above-average rainfall
(4) been polluted by oil
(5) been polluted by acid rain

7. Which of the following statements expresses a fact?

(1) Factories will stop burning coal and oil.
(2) Research could lead to the discovery of a sulfur-free fuel.
(3) Acid rain destroys all living things on contact.
(4) Acid rain is formed by the combining of sulfur oxides with water in the air.
(5) Dangers of acid rain are exaggerated.

8. Which of the following statements can be inferred from the passage?

(1) Sulfur is poisonous to living things.
(2) Plants and animals thrive in water that is clear and blue.
(3) High levels of acidity can be harmful to living things.
(4) Plants and animals die when acid is removed from rainwater.
(5) Plants and animals live best when sulfur is present in the air.

Before you take the GED Mini-Test, check your answers on pages 99–100.

GED Mini-Test 11

TIP Try to answer a question in your mind before reading the GED choices. *Then* locate the choice that is closest to your thoughts.

DIRECTIONS: Choose the one best answer for each item below.

Items 1–4 refer to the following passage.

Most of the energy we use every day comes from fossil fuels. Fossil fuels were formed in the earth millions of years ago when the remains of dead plants and animals were buried beneath layers of mud. The principle fossil fuels are coal, petroleum and natural gas.

Coal is solid fossil fuel. It was the first fossil fuel to be used by industrialized nations. In the United States today, it is burned primarily to produce electric power. Petroleum, or oil, is liquid fossil fuel. Petroleum is presently the leading fuel in the United States and other industrialized nations. Raw petroleum drawn from the earth is called crude oil. The refining of crude oil produces useful products such as gasoline, fuel oil for home heating, kerosene, plastics, synthetic fibers and pharmaceuticals. Natural gas is less dense than liquid petroleum, so it is usually found on top of oil deposits. Natural gas has the advantage of being a "clean-burning" fuel compared to coal and oil, because it produces less air pollution. Some homes use natural gas for cooking and heating.

1. Which of the following phrases describes the origin of fossil fuels?
   (1) produced from synthetic materials in chemical laboratories
   (2) produced from fossilized materials in chemical laboratories
   (3) formed in the earth from the remains of plants and animals
   (4) formed in the earth from pieces of rock
   (5) produced from crude oil in oil refineries

2. According to the passage, air quality would probably improve if
   (1) oil were to replace coal in the production of electric power
   (2) natural gas were to replace oil as the leading fuel
   (3) coal were to replace oil in home heating
   (4) coal were to replace natural gas in home heating
   (5) coal were to replace oil in industry

3. Based on the passage, which of the following statements *best* describes the relationship between fossil fuels and industry?
   (1) Fossil fuels have always been of little importance to industry.
   (2) Fossil fuels were once important, but are now of little importance.
   (3) Fossil fuels were once of little importance, but are now important.
   (4) Fossil fuels have been and continue to be of great importance to industry.
   (5) Fossils fuels are of little importance now but will be of great importance in the future.

4. A shortage of petroleum would probably not affect the availability of materials needed to manufacture which of the following products?
   (1) 100% cotton shirt
   (2) prescription drug
   (3) dress made of synthetic linen
   (4) plastic kitchen utensils
   (5) high-octane gasoline

GO ON TO THE NEXT PAGE.

Items 5–6 refer to the following food chain.

The food chain shows how an organism obtains energy.

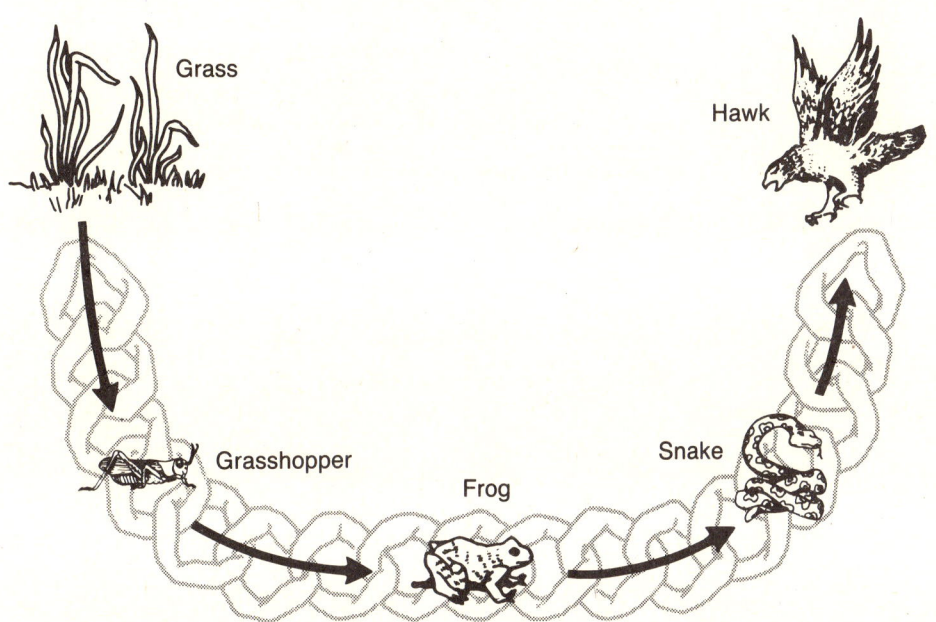

5. Which of the following statements *best* expresses the main idea conveyed by the diagram?

   (1) Frogs are more likely to eat grasshoppers than any other type of insect.
   (2) Hawks are known for stealing food from other animals.
   (3) Living things usually only eat one type of food.
   (4) Living things are dependent upon each other for food.
   (5) Organisms involved in a food chain are likely to become extinct.

6. Which of the following would be the result if the frog were removed from this chain?

   (1) The snake, grasshopper and hawk would lose their food supply.
   (2) Only the snake would lose its food supply.
   (3) The grasshopper and hawk would lose their food supply.
   (4) The snake and hawk would lose their food supply.
   (5) Only the hawk would lose its food supply.

Check your answers to the GED Mini-Test on page 100.

Answers and Explanations

Practice   *pp. 96–97*

1. **Answer:** (4)   Choices (1) and (2) are incorrect because the article makes no mention of size or cost. Choices (3) and (5) are incorrect because both are characteristics of earlier wind generators.

2. **Answer:** (2)   Modern wind generators are described, but their uses are not specified. The other choices are incorrect because these topics are all covered in the article.

3. **Answer:** (3) This choice is clearly stated in the article. The other choices may be true, but they do not relate to the information provided.

5. **Answer:** (4) Choices (1) and (2) are incorrect because oxygen and moisture help form acid rain. Choices (3) and (5) are incorrect because they do not relate to the information given.

7. **Answer:** (4) This statement can be proved. Choices (1), (2) and (5) are incorrect because they are expressions of opinion. Choice (3) is not a fact.

4. **Answer:** (5) Choice (1) is incorrect because energy planners do not expect wind energy to meet all our energy needs. Choice (2) is incorrect because energy planners clearly recommend the use of wind energy. Choice (3) is irrelevant to the present time. Choice (4) is incorrect because wind energy generates electric power.

6. **Answer:** (5) Choice (2) is incorrect because it is a lack of plants and animals that makes the lake look clear and blue. Choices (1), (3) and (4) are incorrect because they do not pertain to information given in the passage.

8. **Answer:** (3) This is correct because both acidic rain and acidic soil damage or kill plants and animals. Choices (1) and (5) are incorrect because no mention is made of the effect of pure sulfur on living things. Choice (2) is incorrect because the only mention of clear, blue water in this passage relates to a lake polluted by acid rain. Choice (4) is incorrect because it is the presence of acid in rainwater that can kill plants and animals.

GED Mini-Test   *pp. 98–99*

1. **Answer:** (3) Choices (1), (2) and (5) are incorrect because fossil fuels were formed naturally in the earth. Choice (4) is incorrect because fossil fuels did not form from pieces of rock.

3. **Answer:** (4) The other choices are incorrect because the passage clearly refers to the involvement of fossil fuels in industry, both in the past and the present.

5. **Answer:** (4) This is correct because each organism in the chain becomes food for another organism. Choice (5) is incorrect because organisms in a food chain are likely *not* to become extinct as long as the food chain holds up. Choices (1), (2) and (3) are incorrect because they totally miss the point of the diagram.

2. **Answer:** (2) All the other choices are incorrect because coal and oil are mentioned together in the passage as producing more air pollution than natural gas.

4. **Answer:** (1) This is correct because cotton is a natural fiber. Choice (2) is incorrect because pharmaceuticals are produced from petroleum. Choice (3) is incorrect because synthetic fibers are produced from petroleum. Choices (4) and (5) are incorrect because plastic and gasoline are produced from petroleum.

6. **Answer:** (4) This is correct because the frog is eaten by the snake who in turn is eaten by the hawk. Choices (1) and (3) are incorrect because the grasshopper is eaten by the frog, not the other way around. Choices (2) and (5) are incorrect because they fail to include the dependency of the hawk on the snake.

12 The Changing Earth

The earth is constantly changing. Some of the changes are slow and gradual. Others are quick and dramatic. In this lesson you will learn what causes some of these changes and how they affect the earth and its people.

Changes in the earth's surface can be caused by mass wasting. **Mass wasting** occurs when the force of gravity pulls rocks and soil down mountain slopes. As sediments come to rest at the bottom of a slope, they form what is called a **talus slope.**

Mass wasting can occur rapidly or slowly. One type of rapid mass wasting is a landslide. During a **landslide,** huge quantities of soil, small stones and large rocks tumble down a mountain. A landslide can be caused by an earthquake, volcanic eruption or heavy rain—any natural event that weakens the supporting rock.

Another type of rapid mass wasting is a mudflow. A **mudflow** is usually caused by a heavy rain. As rain mixes with soil to form mud, gravity begins to pull the mud downhill. As the mud moves, it picks up more soil and becomes thicker. It is difficult to imagine the power of a mudflow—it can move just about anything in its path, including a whole house!

Slow mass wasting can occur in an earthflow. Usually caused by a heavy rain, an **earthflow** consists of the slow, downhill movement of soil and plant life.

Soil creep is the slowest form of mass wasting. Soil particles that have been disturbed by heavy rain, alternate periods of freezing and thawing or animal activity are pulled downhill by gravity. Soil creep is so slow that its effects often go unnoticed for quite some time. An eventual evidence of soil creep is often tilted trees and telephone poles along the side of a steep slope.

Homes devastated by the 1964 Turnagain Heights earthquake and landslide. Anchorage, Alaska.

Strategies for READING

Identify Cause and Effect Relationships

A cause and effect relationship indicates how one thing affects another. A cause is what makes something happen. An effect is what happens as a result.

This entire lesson is about **cause** and **effect.** Every time the earth changes, a particular cause is responsible for the change. The change is the effect, or result, of the cause. The article on the previous page begins with a cause and effect statement. The cause is mass wasting; the effect is changes in the earth's surface.

Sometimes an effect goes on to become the cause of other effects. For example, an earthquake is the effect of certain events in the earth's interior. Once an earthquake occurs, however, it becomes the cause of further effects, such as collapsed buildings and buckled roadways. As you read, it is important to be aware of cause and effect relationships. Remember that words and phrases to identify these relationships include: because, effect, affect, resulted in, occurs when, was caused by, led to, and due to.

Examples

DIRECTIONS: Use the information on this and the preceding page to choose the one best answer for each item below.

1. According to the information provided, which of the following is *not* a cause of mass wasting?

    (1) heavy rain
    (2) animal activity
    (3) earthquakes
    (4) wind
    (5) volcanic eruptions

   Answer: (4)   Choice (1) is incorrect because it can cause all types of mass wasting. Choices (3) and (5) are incorrect because they can cause landslides. Choice (2) is incorrect because it can cause soil creep.

2. A row of tilted trees along a steep slope could be the effect of which of the following?

    (1) a talus slope
    (2) soil creep
    (3) a mudflow
    (4) a landslide
    (5) rapid mass wasting

   Answer: (2)   Choice (1) is incorrect because it is the *result* of mass wasting. Choices (3), (4) and (5) are incorrect because they all cause violent, not gradual, changes.

102   READING: ANALYSIS

Practice

When reading a long passage, underline key words in each paragraph as you read. Then go back and review the passage a second time, using your key words as a guide.

DIRECTIONS: Choose the one best answer for each item below.

Items 1–4 refer to the following article.

An **earthquake** is the shaking and trembling that results from the sudden movement of rock in the earth's crust. When a severe earthquake hits a populated area, there can be tremendous destruction and hundreds of deaths.

The most common cause of earthquakes is faulting. A **fault** is a break in the earth's crust. During faulting, rocks along the fault begin to move. They break and slide past each other. Parts of the earth's crust may be pushed together or pulled apart. During this process, energy is released.

The point beneath the earth's surface where the rocks break and move is called the **focus** of the earthquake. Directly above the focus, on the earth's surface, is the **epicenter.** The most violent shaking occurs at the epicenter.

When rocks in the earth's crust break, waves travel out in all directions from the focus. These waves are known as **seismic waves.** There are three main types of seismic waves.

The seismic waves that travel the fastest are called **primary waves,** or **P waves.** P waves can travel through solids, liquids and gases. P waves are push-pull waves; that is, they cause particles of rock to move back and forth in the same direction as the wave is moving.

The seismic waves that travel the next fastest are **secondary waves,** or **S waves.** S waves can travel through solids but not through liquids or gases. Rock particles disturbed by S waves move from side to side at right angles to the direction the wave is traveling.

The slowest seismic waves are **surface waves,** or **L waves.** L waves travel from the focus directly up to the epicenter. It is the L waves that cause the earth to bend and twist so that whole buildings are swallowed up by the ground.

The more energy an earthquake releases, the stronger and more destructive it is. The strength of an earthquake is measured on a scale called the Richter scale. The **Richter scale** measures how much energy an earthquake releases by assigning the earthquake a number from one to ten. Any number above six on the Richter scale indicates a very destructive earthquake.

GO ON TO THE NEXT PAGE.

1. According to the article, the strength of an earthquake is directly related to which of the following?

   (1) length of the fault
   (2) amount of energy released
   (3) speed of the seismic waves
   (4) distance of the focus from the epicenter
   (5) amount of rock broken

2. Which of the following would *not* be an effect of a severe earthquake?

   (1) the destruction of buildings
   (2) injury or death to human beings
   (3) violent twisting of the earth
   (4) the creation of a fault
   (5) the release of L waves

3. Which of the following statements about seismic waves can be inferred from the article?

   (1) L waves are the most destructive.
   (2) P waves are the most destructive.
   (3) All seismic waves except L waves are destructive.
   (4) All seismic waves are equally destructive.
   (5) Seismic waves are not destructive.

4. Scientists have found that P waves travel through certain areas beneath the earth's crust that S waves cannot. What conclusion could be drawn from this evidence? These areas

   (1) must be solid rock
   (2) must be liquid or gas
   (3) must have very low temperatures
   (4) must never have had earthquakes
   (5) must have had only mild earthquakes

Items 5–6 refer to the following passage.

Erosion is the moving and wearing away of rock materials by natural agents. A dramatic agent of erosion is a glacier.

A **glacier** is a large mass of moving ice. Most glaciers form in mountains where snow accumulates faster than it can melt. As snow falls upon snow, year after year, the snow changes into ice. When the ice becomes heavy enough, the pull of gravity causes it to move down the mountain. As the glacier moves, it picks up large blocks of rock. As the rocks become frozen into the bottom of the glacier, they carve away more rock. Smooth surfaces of the mountain become rough and sharp.

If rock is scooped out of a hollow of a mountain, a bowl-shaped crater called a **cirque** is formed. If several cirques are carved out close to one another, the sharp ridges between the cirques will join to form a peak called a **horn.**

Sometimes, after flowing down a mountain, a glacier will enter a river valley that is narrower than the glacier. As the glacier squeezes through the valley, it erodes both the floor and sides of the valley. As a result, the shape of the valley changes from a V-shaped valley to a broad U-shaped valley.

5. Which of the following ideas is *most* important in defining glaciers?

   (1) heavy snowfall
   (2) heavy rainfall
   (3) erosion
   (4) cirques
   (5) gravity

6. Which of the following is *not* an effect of a moving glacier?

   (1) the formation of cirques
   (2) the formation of a horn
   (3) the smoothing of mountain peaks
   (4) the change of shape of a valley
   (5) the carving away of rock

Before you take the GED Mini-Test, check your answers on pages 106–107.

GED Mini-Test 12

TIP Make up a picture in your mind of what you are reading. This will help you to better understand and remember what you have read. Increasing your understanding will help you pass the GED test.

DIRECTIONS: Choose the one best answer for each item below.

Items 1–4 refer to the following article.

Deep within the earth, there exists hot liquid rock called magma. In some places magma works its way toward the earth's surface by melting solid rock or by moving through cracks in rock. When magma reaches the earth's surface, it is called lava. The place where lava reaches the earth's surface is called a volcano.

In every volcano there is at least one opening called a vent. It is through the vent that the volcano erupts. You may think of a volcanic eruption as being a violent, dramatic event. Sometimes it is, but a volcanic eruption can also be a quiet flow of lava.

Volcanoes can be classified according to the type of volcanic eruptions that form them. There are three main types of volcanoes.

Cinder cone volcanoes are formed from explosive eruptions. Explosive eruptions are caused when lava in vents hardens into rocks. Steam and new lava build up under the rocks, causing pressure. Eventually the pressure becomes great enough to cause a violent explosion. The volcano is formed out of cinders and other rock particles that are blown into the air. Cinder cone volcanoes are characterized by a narrow base and steep sides.

Shield volcanoes result from quiet lava flows. The lava from shield volcanoes flows over a large area because it is thin and runny. A shield volcano, which forms after several quiet eruptions, is a gently sloping, dome-shaped mountain.

Composite volcanoes build up when violent and quiet eruptions alternate. First a violent eruption spews rock and cinders onto the earth. Then a quiet eruption occurs, producing a lava flow that covers the cinders and rock particles. After many alternating eruptions, a cone-shaped mountain is formed. Two very famous composite volcanoes are Mt. Vesuvius in Italy and Mt. Etna in Sicily.

Volcanoes are like "windows" that let us see inside the earth. By analyzing the chemical composition of lava, scientists are able to determine the chemical composition of the magma from which the lava formed.

1. Which of the following is a necessary condition for the formation of a volcano?

    (1) a violent eruption of lava
    (2) the melting of solid rock
    (3) the hardening of lava in vents
    (4) the release of cinders into the air
    (5) the movement of magma to the earth's surface

2. According to the article, the classification of volcanoes is based on which of the following?

    (1) their shape
    (2) their size
    (3) the number of vents they have
    (4) the events that cause them to form
    (5) the effect they have on the environment

GO ON TO THE NEXT PAGE.

3. The formation of a gently sloping, dome-shaped volcano is the result of which of the following events?

   (1) several violent eruptions
   (2) alternating violent and quiet eruptions
   (3) several quiet eruptions
   (4) the release of cinders into the air
   (5) the formation of steam in vents

4. Which of the following statements about lava can be inferred from the article?

   (1) Lava produces steam.
   (2) Lava can be thin and runny.
   (3) Lava always flows over a wide area.
   (4) Lava produces magma.
   (5) Lava consists primarily of cinders.

Items 5–6 refer to the following map.

This map shows the world zones of earthquake and volcano activity. Most major earthquakes and volcanic eruptions occur in these zones.

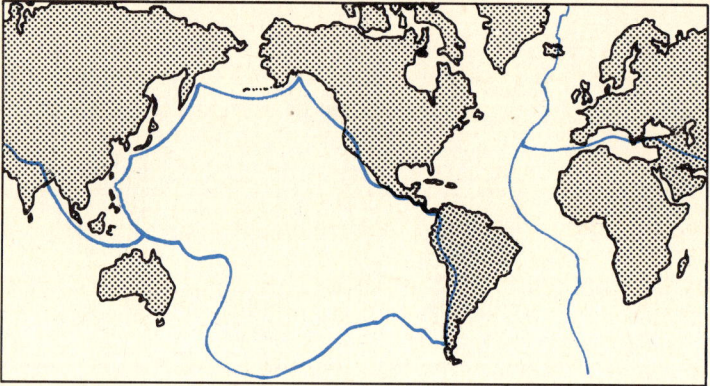

5. According to the map, there is evidence to support which hypothesis?

   (1) Areas close to oceans and seas are more likely to have serious earthquakes than inland areas.
   (2) Earthquakes do not occur on islands.
   (3) Earthquakes are more likely to occur in the northern hemisphere than in the southern hemisphere.
   (4) Earthquakes are more severe in the southern hemisphere than in the northern hemisphere.
   (5) Earthquakes are more likely to occur on small than on large continents.

6. Which of the following areas would provide the *best* opportunity to study volcanic activity?

   (1) northeastern Canada
   (2) western South America
   (3) western Africa
   (4) Australia
   (5) eastern South America

Check your answers to the GED Mini-Test on page 107.

Answers and Explanations

Practice pp. 103–104

1. **Answer:** (2) Choices (1), (3) and (4) are incorrect because, while they may be significant in other ways, they do not relate to the strength of an earthquake. Choice (5) may relate to the strength of an earthquake, but this is not mentioned in the article.

2. **Answer:** (4) This is correct because faulting *causes* an earthquake. This question is a good example of how answers can reverse the cause and effect of a situation and thereby tempt you to choose an attractive, but incorrect, answer.

3. **Answer:** (1) This is correct since it is the L waves that cause the upheavals in the ground that are so destructive. Choices (2) and (3) are incorrect because they contradict the correct answer. Choices (4) and (5) are incorrect because the article emphasizes the destruction that results from L waves, but not from any other waves.

4. **Answer:** (2) This is correct because while P waves travel through solids, liquids and gases, S waves travel through only solids. Thus, choice (1) is incorrect. Choices (3), (4) and (5) are incorrect because the article does not relate seismic waves to these factors.

5. **Answer:** (5) Movement that is caused by gravity is part of the definition of glaciers. Choice (1) does not include the accumulation of snow and ice. Choices (3) and (4) are effects of moving glaciers. Choice (2) is irrelevant.

6. **Answer:** (3) This is correct because a glacier does the opposite—it makes smooth peaks jagged. All the other choices are effects of a moving glacier.

GED Mini-Test pp. 105–106

1. **Answer:** (5) The key word in this question is necessary. All the other choices can be associated with the formation of certain volcanoes, but only choice (5) is a necessary condition for the formation of *all* volcanoes.

2. **Answer:** (4) Choices (1), (2), (3) and (5) are incorrect because, although these factors are alluded to in the article, they are not distinctive enough in each type of volcano to be considered a basis for classification.

3. **Answer:** (3) Choices (1) and (2) are incorrect because they form cone-shaped volcanoes. Choices (4) and (5) are incorrect because they are associated with the formation of cone-shaped volcanoes.

4. **Answer:** (2) Choices (1) and (5) are incorrect because the article does not link these factors to lava in cause and effect relationships. Choice (3) is incorrect because lava does not flow over a wide area in a violent eruption. Choice (4) is incorrect because lava is formed from magma, not the other way around.

5. **Answer:** (1) Most of the land areas covered by the earthquake and volcano zones are islands, coastal areas or peninsulas. Choice (2) is incorrect because it contradicts the correct response. Choice (3) is incorrect because the zones cover about equal areas in both hemispheres. Choice (4) cannot be determined from the information given. Choice (5) is incorrect because it is actually the larger continents that tend to have more earthquake and volcano zones.

6. **Answer:** (2) Choices (1), (3), (4) and (5) are incorrect because the zones shown on the map do not pass over these areas.

REVIEW
Earth Science

In this section you have read about the earth's structure and history. You have learned about the earth's atmosphere, oceans and natural resources. You have also studied some of the ways in which the earth changes, both above and below the surface.

The following exercises summarize and expand upon these concepts. As you answer each set of questions, watch for words and ideas that are familiar to you.

DIRECTIONS: Choose the <u>one</u> best answer for each item below.

Items 1–4 refer to the following diagram.

The diagram below summarizes the major characteristics of the layers of the earth.

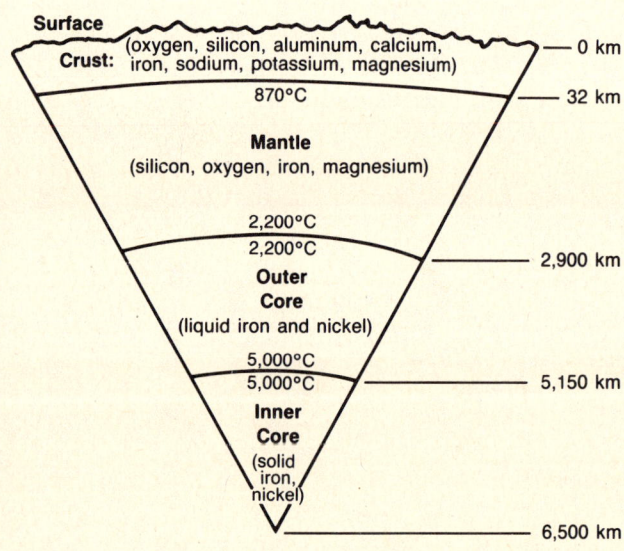

1. Which statement correctly describes the temperature of the earth as one moves from the surface to the inner core?

   (1) The temperature decreases steadily.
   (2) The temperature increases steadily.
   (3) The temperature increases, then decreases.
   (4) The temperature decreases, then increases.
   (5) The temperature remains constant.

2. Which of the following elements is present in all four layers of the earth?

   (1) oxygen
   (2) silicon
   (3) aluminum
   (4) iron
   (5) nickel

3. Compared to the rest of the earth, the depth of the crust is

   (1) more than 50%
   (2) about 50%
   (3) about 25%
   (4) about 10%
   (5) less than 1%

4. According to the diagram, which layer of earth has the *most* varied composition?

   (1) inner core
   (2) outer core
   (3) mantle
   (4) crust
   (5) all have similar composition

GO ON TO THE NEXT PAGE.

Items 5–6 refer to the following diagram.

The diagram below shows the characteristics of the four main layers of the atmosphere.

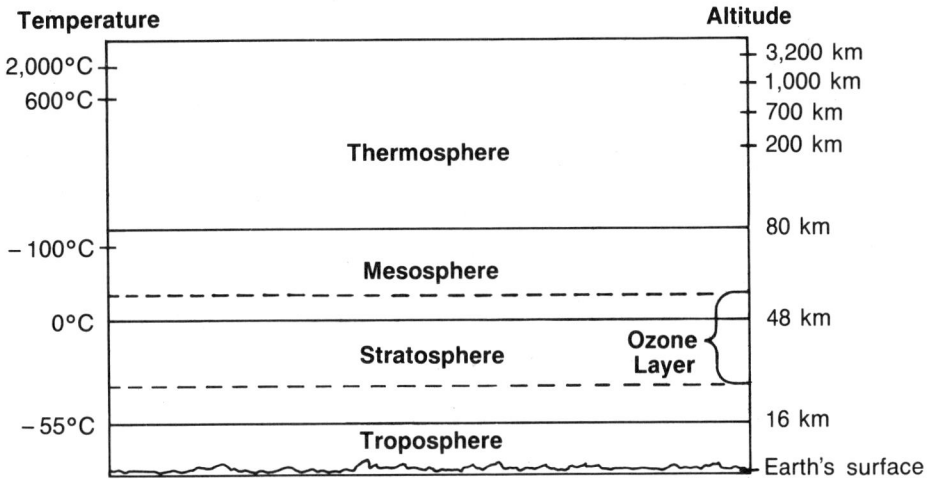

5. According to the diagram, temperature in the earth's atmosphere

   (1) remains essentially constant as altitude increases
   (2) increases as altitude increases
   (3) decreases as altitude increases
   (4) increases, then decreases as altitude increases
   (5) decreases, then increases, then decreases, then increases as altitude increases

6. The upper thermosphere is called the exosphere, where artificial satellites orbit the earth. According to the diagram, which statement *must* be true about these satellites?

   (1) They orbit the earth at an altitude of 100 to 200 kilometers.
   (2) They are protected from ultraviolet radiation by the ozone layer.
   (3) They can withstand extremely high temperatures.
   (4) They pass through thin clouds of ice.
   (5) They play a role in television transmission.

Items 7–8 refer to the following diagram.

Land is one of the earth's most important resources. The diagram shows the condition of land in relation to its usefulness for farming.

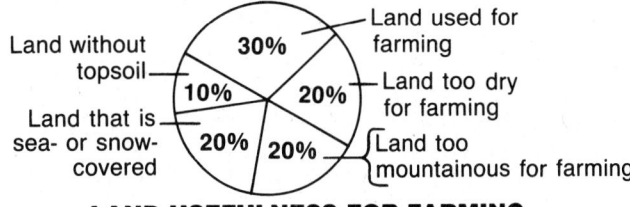

LAND USEFULNESS FOR FARMING

7. According to the graph, which of the following is *not* a problem in finding suitable land for farming?

   (1) too much snowfall
   (2) too much rainfall
   (3) land too mountainous
   (4) land under water
   (5) lack of topsoil

8. Irrigation brings water to dry land. If all of the land too dry for farming were sufficiently irrigated, what percentage of available land would be suitable for farming?

   (1) 10%
   (2) 20%
   (3) 30%
   (4) 50%
   (5) 100%

Check your answers to the Review on page 156.

OVERVIEW
Chemistry

An ice crystal painting formed by the freezing of water vapor on a window pane.

matter
anything that has mass and takes up space

Chemistry is the study of matter and changes in matter. You may think that chemistry exists only in a laboratory, but chemistry is all around you and within you. In fact it is a chemical reaction within your brain that is enabling you to understand and remember these words as you read them!

In the first lesson of this section you will learn about matter. **Matter** is what everything in the world is made up of. Matter can exist in any one of three common physical states—solid, liquid or gas.

The building blocks of matter are tiny particles called **atoms.** Atoms contain three important subatomic particles called protons, neutrons and electrons. Atoms bond together chemically to form molecules. A **molecule** is the smallest particle of a compound that still has all the properties of that compound.

In this lesson you will also learn about the elements. An **element** is the simplest of substances, for it cannot be broken down into other substances by chemical means. Scientists have listed the elements in an arrangement called the **Periodic Table.** As you study the Periodic Table, you will find out how useful it is in predicting the properties of elements.

Chemical reactions involve the changing of substances into new and different substances. In the second lesson of this section you will discover how chemical changes take place. You will learn that during chemical reactions bonds between atoms are broken and new bonds are formed. You will learn how energy changes accompany chemical reactions, and you will also learn about a very interesting aspect of chemical change called **entropy.**

In this lesson you will also learn about a familiar type of mixture called a solution. A **solution** consists of a substance called a solute dissolved in another substance called the solvent. Solutions can be made up of solids, liquids or gases. Many of the most familiar solutions are solids, liquids or gases dissolved in a liquid solvent.

In this lesson you will learn about an important group of chemical compounds known as acids, bases and salts. Some of these compounds are very familiar to you. You can find an **acid** in an orange, a **base** in soap and cleaning solutions and a **salt** right on your dinner table. And speaking of dinner, the last part of this lesson will teach you how to become a "kitchen chemist"—as you learn what really happens when baking powder makes bread dough rise!

atom
the smallest particle of an element that has all the properties of that element

molecule
the smallest particle of a compound that has all the properties of that compound

element
a substance that cannot be broken down into simpler substances by chemical means

Periodic Table
an arrangement of the elements in order of increasing atomic number

entropy
a measure of the degree of disorder in a substance or system

solution
a homogeneous mixture of two or more substances dissolved in each other

acid
substance that releases hydrogen ions in water solution

base
substance that releases hydroxide ions in water solution

salt
a neutral compound that results from the chemical combination of an acid and a base

Matter

Matter is anything that has mass and takes up space. Matter is what the world is made of. You can better understand the definition of matter if you think of an everyday object, such as a can of soup. If you were to put the can of soup on a balance, the scale would tip—so you know that the can of soup has mass. If you were to line up many cans of soup on a shelf, you would eventually reach a point where no more cans would fit—so you know that a can of soup takes up space.

Matter can exist in any of three common physical states. These states are solid, liquid and gas. A **solid** is any form of matter that has definite shape and volume. A block of wood is an example of a solid. If you try to put a square block of wood into a round hole, it simply will not fit. The wood has definite shape. Similarly, if you try to put the block of wood into a space that is too small, it also will not fit. The block of wood has definite volume. A **liquid** has definite volume, but it does not have definite shape. If you pour a quart of milk into a gallon jug, the milk will fill only one-quarter of the jug. If you pour the same quart of milk into an eight-ounce glass, the milk will overflow. The volume of the milk does not change. The shape of a liquid, however, will change each time you pour it into a different container. A **gas** has neither definite shape nor definite volume. Perhaps you have seen party balloons made in the shapes of animals or cartoon characters. The air or helium inside the balloon simply fills the shape of the balloon. A gas will also spread out to fill the volume of a container that it is placed in. You can understand this property of a gas if you think of how quickly the smell of an apple pie baking in the oven will fill the whole kitchen.

Most matter can change from one state to another. Such a change is called a **phase change.** If enough heat is removed from a liquid, it will freeze into a solid. Similarly, if enough heat is added to a solid, it will melt into a liquid. The temperature at which these phase changes occur is called the **freezing point** or **melting point** of a substance. If enough heat is added to a liquid, the liquid will change into a gas. This process is called **vaporization.** The temperature at which vaporization occurs is called the **boiling point** of a substance. When a gas is cooled sufficiently to change into a liquid, the process is called **condensation.**

INTRODUCTION

Strategies for READING

Use Given Ideas in Another Context

This skill involves taking information you already know and making it work for you in a new situation.

Many of the things you will read in science include the explanation of a definition or general principle. By applying the theoretical principle to concrete situations in another **context,** they become easier to understand.

In the article on the previous page, the author uses the example of a can of soup to help you understand the definition of matter. The author shows how a can of soup can be "tested" for the two properties that characterize matter. Try to picture the soup can tipping a scale and taking up space on a shelf. A general principle will seem more real to you if you can visualize its application.

Examples

DIRECTIONS: Use the information on this and the preceding page to choose the <u>one</u> best answer for each item below.

1. Which of the following situations would produce a phase change?

   (1) A drop of food coloring is added to a glass of water.
   (2) A container of ice cream is left on top of a radiator.
   (3) A hole is pricked in a balloon.
   (4) A can of soup is placed on a scale.
   (5) A square block of wood is cut in half.

Answer: (2) This is correct because the solid ice cream would melt into a liquid. Choices (1), (3) and (5) are incorrect because changing color, volume or size does not indicate a phase change. Choice (4) is irrelevant.

2. A heart-shaped cake is made by pouring cake batter into a heart-shaped mold. Which principle illustrates this situation?

   (1) A solid conforms to the shape of the container into which it is placed.
   (2) A solid has definite shape but not definite volume.
   (3) Liquids have definite shape and volume.
   (4) Liquids have no definite shape.
   (5) Gases have definite shape and volume.

Answer: (4) This is correct because liquid cake batter would conform to the shape of the mold. Choices (1), (2), (3) and (5) are untrue.

READING: APPLICATION   113

Practice

A key to applying the reading skill **use given ideas in another context** is to apply the theoretical information to a concrete situation that is familiar to you. Then you can transfer this to another context.

DIRECTIONS: Choose the one best answer for each item below.

Items 1–4 refer to the following passage.

All matter is made up of elements. An **element** is a substance that cannot be broken down into simpler substances by chemical means. Oxygen, carbon, iron and aluminum are some common elements. Water is not an element because it can be broken down into the elements hydrogen and oxygen. An interesting question to consider is, can a sample of an element be broken down into smaller and smaller parts forever and still be an element? Can, for example, a piece of copper wire be cut into smaller and smaller pieces and still be copper?

The first person to come up with an answer to this question was the Greek philosopher Democritus. Over 2,000 years ago, Democritus stated that if a substance were divided into smaller and smaller pieces, eventually a smallest piece would be obtained that could no longer be divided. Democritus called this smallest piece "atom," from the word *atomos*, which means "cannot be divided." Although Democritus did not understand matter the way modern scientists do, his basic idea was remarkably correct in light of modern atomic theory.

Today scientists know that an **atom** is the smallest particle of an element that still has the properties of that element. All elements are made up of atoms. Atoms are so small that scientists have never seen them, not even with the most powerful microscopes. But, by carrying out many experiments, scientists have been able to determine a great deal about what atoms must be like.

Every atom consists of a positively charged nucleus surrounded by negatively charged electrons. The nucleus, which accounts for most of the atom's mass, contains positively charged particles called **protons.** The nucleus also contains particles with no charge called **neutrons.** A proton and a neutron have approximately the same mass.

Revolving around the nucleus are tiny, negatively charged particles called **electrons.** Electrons are much, much smaller than protons or neutrons. Electrons occupy a certain region around the nucleus. This region is often referred to as an **electron** cloud.

GO ON TO THE NEXT PAGE.

114 PRACTICE

1. Which of the following statements about substance X would indicate that it is *not* an element?

   (1) It decomposes when heated to produce mercury and oxygen.
   (2) It is red-orange in color.
   (3) It reacts with water to form a base.
   (4) It is a solid at room temperature.
   (5) It can react with certain compounds to form salts.

2. Which of the following *best* describes Democritus' contribution to modern science?

   (1) He saw atoms for the first time.
   (2) He measured the size of an atom.
   (3) He determined that atoms contain protons, neutrons and electrons.
   (4) He proposed the idea that is the basis of modern atomic theory.
   (5) He proposed the idea that an element cannot be broken down.

3. The total charge on an atom is zero; an atom is electrically neutral. Which conclusion supports this information?

   (1) The number of protons must be greater than the number of electrons.
   (2) The number of protons must be less than the number of electrons.
   (3) The number of protons must be equal to the number of electrons.
   (4) The number of protons must be equal to the number of neutrons.
   (5) The number of neutrons must be equal to the number of electrons.

4. Atom X has 5 protons more than atom Y. Both atoms have the same mass. Which statement explains why this is so?

   (1) Atom Y has more electrons than atom X.
   (2) Atom X has more electrons than atom Y.
   (3) Atom Y has more neutrons than atom X.
   (4) Atom X has more neutrons than atom Y.
   (5) Atom X has the same number of neutrons as atom Y.

Items 5–6 refer to the following passage.

When two or more elements combine chemically, the result is a **compound.** Water is a compound, because it consists of the elements oxygen and hydrogen.

The smallest particle of a compound that still has the properties of that compound is called a **molecule.** A molecule is made up of two or more atoms chemically bonded together. When atoms bond together, they share, or transfer, electrons. If the electrons are shared, the bond is said to be **covalent.** If the electrons are transferred from one atom to another, the bond is said to be **ionic.** Some atoms form bonds easily with other atoms, while some atoms hardly ever form bonds.

5. Which of the following combinations represents a compound?

   (1) two carbon atoms side-by-side in a diamond crystal
   (2) sugar and water mixed together
   (3) a hydrogen atom and a chlorine atom sharing a pair of electrons
   (4) hydrogen and nitrogen existing together in air
   (5) water and carbon dioxide mixed together in carbonated water

6. It can be inferred from the passage that atoms that do not bond easily with other atoms

   (1) have no electrons
   (2) have little molecular activity
   (3) are present in very few compounds
   (4) are found only in water
   (5) form only covalent bonds

Before you take the GED Mini-Test, check your answers on page 118.

GED Mini-Test 13

TIP: Try to spot wrong answers among the answer choices. Often the choices reflect common errors. Look for them and eliminate them. Then locate the correct answer.

DIRECTIONS: Choose the one best answer for each item below.

Item 1–6 refer to the following passage and the diagram on page 117.

The atoms of each element are different from the atoms of any other element. What accounts for this difference? It is the number of protons in the nucleus that makes atoms different from other atoms. The number of protons in the nucleus of an atom of an element is called the atomic number of that element. Each element has a unique atomic number.

Scientists have found that when the elements are arranged in order of increasing atomic number, their properties vary in a regular way. This observation is stated in the Periodic Law: *The physical and chemical properties of the elements are periodic functions of their atomic numbers.*

Based on the periodic law, scientists have developed an arrangement of the elements called the Periodic Table. In the Periodic Table elements are listed in order of increasing atomic number. The horizontal rows of the Periodic Table are called periods. The vertical columns of the table are called groups or families.

Because the properties of the elements vary in a regular pattern, you can tell a lot about an element just by where it is placed in the Periodic Table. For example, all of the elements on the right side of the Periodic Table are nonmetals. Elements on the left side and in the center of the table are metals. The solid zig-zag line that you see near the right side of the table is the dividing line between metals and nonmetals.

Elements in the same group or family have similar properties. For example, all of the elements in group I, with the exception of hydrogen, are very active metals. All of the elements in group II are fairly active metals that are solids at room temperature. The elements in group VII are all very active nonmetals that combine readily with group I metals to form salts. And the elements in group VIII are all very inactive gases.

GO ON TO THE NEXT PAGE.

1. According to the passage, the Periodic Table is based on a relationship between which of the following two factors?

   (1) atomic number and atomic size
   (2) atomic number and number of electrons
   (3) atomic number and physical and chemical properties
   (4) number of electrons and physical and chemical properties
   (5) atomic size and metallic properties

2. It can be inferred from the passage that the word "periodic" means

   (1) arranged from smallest to largest
   (2) arranged randomly
   (3) existing in a unique state
   (4) having similar properties
   (5) having similar properties appear at regular intervals

3. According to the information provided, the first member of period 3 must be

   (1) a very active metal
   (2) a fairly active metal
   (3) a very inactive metal
   (4) a very active nonmetal
   (5) a very inactive gas

4. According to the Periodic Table, which of the following three elements would have similar properties?

   (1) sodium (Na), chlorine (Cl), hydrogen (H)
   (2) neon (Ne), argon (Ar), krypton (Kr)
   (3) lithium (Li), magnesium (Mg), sulfur (S)
   (4) oxygen (O), carbon (C), chlorine (Cl)
   (5) phosphorus (P), sulfur (S), chlorine (Cl)

5. According to the Periodic Table, which of the following elements must be a metal?

   (1) xenon (Xe)
   (2) radon (Rn)
   (3) selenium (Se)
   (4) palladium (Pd)
   (5) nitrogen (N)

6. According to the information provided, the element iodine (I) would combine with which element to form a salt?

   (1) carbon (C)
   (2) potassium (K)
   (3) boron (B)
   (4) bromine (Br)
   (5) xenon (Xe)

Check your answers to the GED Mini-Test on page 118.

Answers and Explanations

Practice pp. 114–115

1. **Answer:** (1) This is correct because an element cannot be broken down into other elements. Choice (2) is incorrect because nothing stated in the passage restricts the color of an element. Choices (3) and (5) are incorrect because elements can react with other substances. Choice (4) is incorrect because many elements are solids.

2. **Answer:** (4) Choices (1) and (2) are incorrect because no one has ever been able to see or directly measure an atom. Choice (3) is incorrect because this information is part of modern atomic theory. Choice (5) is incorrect because Democritus did not theorize about the definition of an element.

3. **Answer:** (3) Choice (1) is incorrect because the atom would have a net positive charge. Choice (2) is incorrect because the atom would have a net negative charge. Choices (4) and (5) are incorrect because neutrons do not have any effect on the charge of an atom.

4. **Answer:** (3) This is correct because neutrons and protons have the same mass—thus extra neutrons in Y could balance the extra protons in X. Choices (1) and (2) are incorrect because electrons make up so little of an atom's mass that they could not possibly balance the extra protons. Choice (4) would make atom X even heavier. Choice (5) is incorrect because atom X would still be heavier by five protons.

5. **Answer:** (3) This is correct because it describes a covalent bond between hydrogen and chlorine atoms. Choice (1) is incorrect because the carbon is still just one element. Choices (2), (4) and (5) are incorrect because they represent mixtures.

6. **Answer:** (3) Choice (1) is incorrect because it is apparent from the information given that all atoms have electrons. Choice (2) is incorrect because molecules are atomically bonded by definition. Choices (4) and (5) do not relate to the information given.

GED Mini-Test pp. 116–117

1. **Answer:** (3) Although atomic size is related to atomic number, this relationship is not the basis of the Periodic Table—thus choice (1) is incorrect. Choices (2) and (4) are incorrect because these relationships, though valid, are also not the basis of the Periodic Table. Choice (5) is irrelevant.

2. **Answer:** (5) Choice (1) is incorrect because arranging atomic numbers from smallest to largest *results* in periodicity. Choice (2) is incorrect because the Periodic Table is obviously not a random arrangement. Choice (3) is a word taken out of context. Choice (4) is incorrect because elements with similar properties are grouped together as a result of periodicity.

3. **Answer:** (1) This is correct because the first member of any period (horizontal row) must be in group I—and all members of group I below period one are very active metals. Choices (2) through (5) are incorrect because these are not properties of elements in group I.

4. **Answer:** (2) This is correct because these elements are all in the same family, group VIII. The elements in choices (1), (3) and (4) are not all members of the same family. If you answered choice (5), you confused elements in the same period with elements in the same group.

5. **Answer:** (4) Choices (1), (2), (3) and (5) are incorrect because all of these elements are to the right of the metal-nonmetal dividing line.

6. **Answer:** (2) This is correct because elements in group VII, of which iodine is a member, combine with active metals in group I to form salts. Choices (1), (3), (4) and (5) are not group I metals.

14 Chemical Reactions

Chemical reactions involve the breaking and forming of bonds between atoms. Energy is absorbed in the breaking of chemical bonds and released in the formation of new bonds. Usually the net absorption or the net release of energy is in the form of heat.

A chemical reaction acquiring or absorbing heat energy from the surroundings is called an **endothermic reaction.** More heat is absorbed and less heat is released in this type of reaction.

On the other hand, a reaction that releases heat energy to the surroundings is termed an **exothermic reaction.** Less heat is absorbed and more heat is released in this situation. Some reactions of this type can release energy in the form of light or an electric charge.

It is interesting to consider the question of why some chemical reactions occur spontaneously while others do not. In general, reactions occur spontaneously when they involve a change from higher energy to lower energy. Because exothermic reactions move from a state of higher energy to a state of lower energy, they usually occur spontaneously. However, there is another factor besides energy change that determines whether or not a reaction will be spontaneous.

This other factor is called entropy. **Entropy** is a measure of the disorder or randomness in a substance or system. High entropy means a high degree of disorder. A common example of increased entropy is when a substance changes from a liquid to a gas. The molecules in a gas have much greater freedom than the molecules in a liquid, and many more ways in which they can move. Spontaneous reactions tend to be ones that go in the direction of increased entropy.

If a reaction involves both a decrease in energy and an increase in entropy, the reaction will definitely be spontaneous. If a reaction involves an increase in energy and a decrease in entropy, it can never be spontaneous. If, however, one factor in the reaction favors spontaneity while the other does not, the spontaneity of the reaction will depend upon which factor is greater.

Endothermic reaction

(change from liquid to gas)

Exothermic reaction

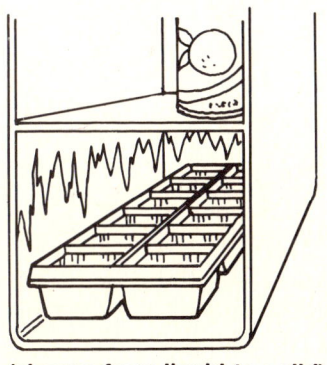

(change from liquid to solid)

INTRODUCTION 119

Strategies for READING

Assess Adequacy of Data to Support Conclusions

This skill may involve applying information in charts, graphs or a written passage to prove or disprove statements.

To assess the accuracy and adequacy of facts, follow these steps:
1. Look for supporting details in the text. Support may be in the form of a chart, table, graph, picture or in words.
2. Make a list of these details for your reference.
3. Study the internal logic and the accuracy of the statement. Ask yourself, Does the statement make sense? Is it logical? Is it correct and accurate? Does the data support the conclusion? If yes, then the data is adequate. If no, then the data is insufficient or the conclusion is inaccurate.

Look at page 119. A conclusion states that heat is released during an exothermic reaction. Assess your information. The statement makes sense, and the data in paragraph two is sufficient to support the conclusion. Suppose a conclusion states that 14.5° of heat is released when a lightning bug glows. Does the picture show this? Is the data adequate to support the conclusion?

Example

DIRECTIONS: Use the information on this and the preceding page to choose the one best answer for the item below.

1. Which of the following conclusions about a specific exothermic reaction could *not* be supported by the information?

   (1) Chemical bonds are formed during the reaction.
   (2) Energy in the form of light is released by the reaction.
   (3) Depending upon the change in entropy, the reaction may occur spontaneously.
   (4) The reaction involves a net decrease in energy.
   (5) More energy is released by the reaction than is absorbed.

Answer: (2) This is correct because the article does not state the specific form of energy released during specific exothermic reactions. Choices (1), (4) and (5) are incorrect because the article includes these conditions to define an exothermic reaction. Choice (3) is incorrect because the article states that most exothermic reactions are spontaneous, but that the determining factor can be the change in entropy.

Practice

HINT Do not skip over a test item just because the subject matter is unfamiliar. The information you need to answer each question is provided by the passage or the illustration that accompanies it.

DIRECTIONS: Choose the <u>one</u> best answer for each item below.

Items 1–2 refer to the following passage.

A **solution** is a homogeneous mixture in which two or more elements or compounds are dissolved in one another. The substance being dissolved is called the **solute.** The substance that does the dissolving is called the **solvent.** In a sugar-water solution, sugar is the solute and water is the solvent.

Solutions can involve solids, liquids or gases. For example, carbonated water, which is carbon dioxide dissolved in water, is an example of a gas-in-liquid solution. The maximum amount of a solute that will dissolve in a given amount of solvent at a certain temperature is called **solubility**. Usually, solubility increases with temperature. For gases dissolved in a liquid, however, the reverse is true—solubility increases as the temperature of the solvent decreases. The rate of solution is affected by temperature. An increase of temperature causes molecules to move and spread apart more quickly.

1. A glass of carbonated beverage left in a warm room goes flat, while a glass of the same beverage stored in the refrigerator does not. According to the information provided, which of the following statements could explain why this happens?

   (1) Water molecules evaporate more rapidly at high temperatures.
   (2) The solubility of a gas in water decreases as temperature increases.
   (3) Molecules move more slowly at low temperatures.
   (4) The solubility of a gas in water decreases as temperature decreases.
   (5) The rate of solution is slower at low temperatures than at high temperatures.

2. The temperature of ocean water drops rapidly from the surface to depths below 300 meters. Which of the following conclusions can be supported by the information provided?

   (1) Water near the bottom of the ocean contains less oxygen than water near the surface.
   (2) Molecules near the bottom of the ocean move faster than molecules near the surface.
   (3) Water from the ocean bottom would be purer than water from the surface.
   (4) Water from the bottom of the ocean is less dense than water from the surface of the ocean.
   (5) Water from the bottom of the ocean would taste less salty than water from the surface.

GO ON TO THE NEXT PAGE.

Items 3–6 refer to the following passage.

Three important groups of chemical compounds are acids, bases and salts. These compounds produce ions when dissolved in water. **Acids** and **bases** can be identified by the type of ions they produce. Acids dissolved in water produce hydrogen, or H^+ ions, while bases produce hydroxide, or OH^- ions. There are many examples of acids and bases in the things we use every day. Citric acid is found in citrus fruits, and acetic acid gives vinegar its sour taste. Ammonium hydroxide, a common cleaning solution, is a base. Another base, magnesium hydroxide, is the active ingredient in many stomach remedies.

A strong acid such as sulfuric acid or a strong base such as sodium hydroxide is poisonous and extremely corrosive. Yet a weak acid such as citric acid or a weak base such as magnesium hydroxide can be safely handled and ingested. The strength of an acid or base is measured on a scale called the pH scale. The **pH scale** consists of numbers from 0 to 14. The number 7 is the neutral point. Pure water has a pH of 7. Substances with a pH above 7 are basic, and substances with a pH below 7 are acidic. The strongest acid would have a pH of 0, and the strongest base would have a pH of 14. When an acid and a base combine chemically, the result is a neutral compound called a **salt**. Water is also a product of this reaction. The most familiar salt is sodium chloride (table salt).

3. Which of the following statements would indicate that substance X is *not* an acid?

   (1) It combines chemically with certain other compounds to produce salts.
   (2) It dissolves in water to produce OH^- ions.
   (3) It can be poisonous if swallowed.
   (4) Its pH is lower than that of sodium hydroxide.
   (5) It can be corrosive to skin.

4. Which of the following represents a correct ordering of substances from lowest to highest pH?

   (1) magnesium hydroxide, pure water, sulfuric acid, sodium hydroxide
   (2) pure water, sulfuric acid, citric acid, sodium hydroxide, magnesium hydroxide
   (3) sulfuric acid, sodium hydroxide, citric acid, magnesium hydroxide
   (4) sulfuric acid, citric acid, pure water, magnesium hydroxide, sodium hydroxide
   (5) sodium hydroxide, magnesium hydroxide, pure water, citric acid, sulfuric acid

5. An antacid relieves indigestion by neutralizing excess acid in the stomach. Which of the following statements *must* be true about an antacid?

   (1) It contains salt.
   (2) It has a low pH.
   (3) It contains a base.
   (4) It produces H^+ ions.
   (5) It dissolves very rapidly.

6. It can be inferred from the passage that a solution of table salt and water would have a pH of approximately

   (1) 0
   (2) 4
   (3) 7
   (4) 10
   (5) 14

GED Mini-Test 14

TIP Use all your test time, even if you finish early. Review your work. Be careful, though, about changing your answers. Make sure you answer *all* the questions and correctly mark your answer sheet.

DIRECTIONS: Choose the one best answer for each item below.

Items 1–4 refer to the following passage.

You may not think of the kitchen as being a chemistry lab, but many chemicals can be found right on your kitchen shelf. One substance that contains several interesting chemicals is baking powder, which is used to make bread and cake dough rise.

The principal ingredient in baking powder is sodium bicarbonate, $NaHCO_3$. When sodium bicarbonate reacts with an acid, it produces the gas carbon dioxide (CO_2) and water. When sodium bicarbonate is heated strongly, it decomposes to form carbon dioxide and sodium carbonate (Na_2CO_3). Baking powder also contains a substance that will react with water to form acids. This substance is usually a type of compound called a tartrate.

1. It can be inferred from the passage that the purpose of the tartrate in baking powder is to

   (1) provide an acid for sodium bicarbonate to react with
   (2) decompose to form carbon dioxide
   (3) react with carbon dioxide
   (4) provide a salt that will make dough rise
   (5) react with water to form sodium bicarbonate

2. When baking powder is added to bread dough, the substance that actually makes the dough rise is

   (1) salt
   (2) oxygen
   (3) water
   (4) tartrate
   (5) carbon dioxide

3. Sodium bicarbonate reacts with vinegar to produce carbon dioxide and water. Which of the following *must* be true about vinegar?

   (1) It contains sodium carbonate.
   (2) It makes bread dough rise.
   (3) It decomposes when heated.
   (4) It contains an acid.
   (5) It contains a salt.

4. When baking powder is left uncovered in damp or humid weather, it quickly loses its effectiveness. Based on the passage, which of the following statements could explain why this happens?

   (1) Moisture in the air causes sodium bicarbonate to decompose.
   (2) Oxygen in the air causes the tartrate to decompose.
   (3) Moisture in the air reacts with the tartrate.
   (4) Oxygen in the air reacts with carbon dioxide.
   (5) Oxygen in the air reacts with sodium bicarbonate.

GO ON TO THE NEXT PAGE.

Items 5–6 refer to the following diagram.

Sodium bicarbonate, the principal ingredient in baking powder, is one of the products of an industrial procedure called the Solvay Process. The steps in the Solvay Process that produce sodium bicarbonate are shown below.

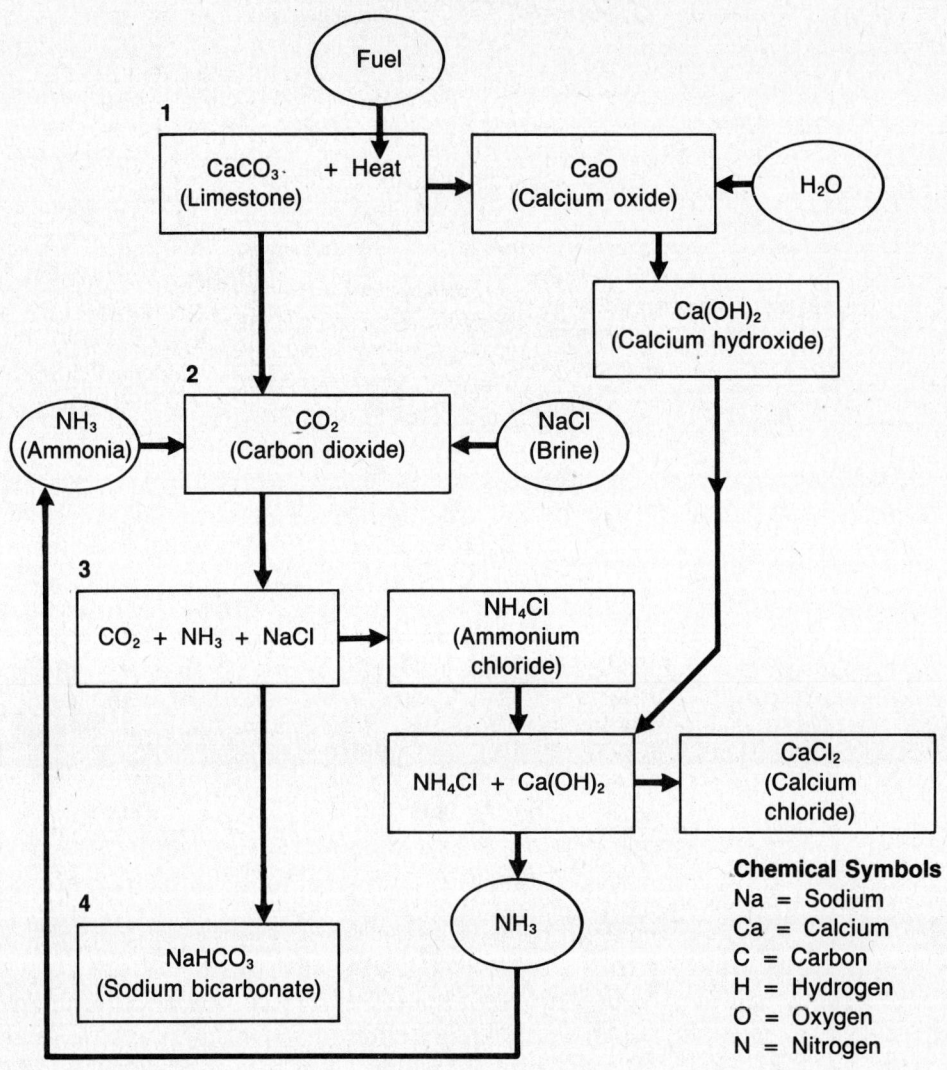

5. According to the diagram, the sodium in sodium bicarbonate comes from which of the following?

   (1) ammonia
   (2) brine
   (3) calcium hydroxide
   (4) limestone
   (5) ammonium chloride

6. If the above reaction is run many times in a row, which of the following substances would *not* have to be added each time?

   (1) limestone
   (2) water
   (3) ammonia
   (4) brine
   (5) fuel

Answers and Explanations

Practice pp. 121–122

1. **Answer:** (2) Choices (1), (3) and (5) may be valid statements, but they do not relate to the problem at hand, nor are they all covered in the passage. Choice (4) is incorrect because it is an untrue statement.

2. **Answer:** (5) This is correct because solids are less soluble in water at low temperatures than at high temperatures. Choice (1) is incorrect because more gas will dissolve in water at low temperatures than at high temperatures. Choices (2), (3) and (4) could be true, but they are not directly discussed in the passage.

3. **Answer:** (2) Choice (1) is incorrect because acids combine with bases to produce salts. Choices (3) and (5) are incorrect because these properties are true if the acid is strong. Choice (4) is incorrect because all acids have a pH lower than that of sodium hydroxide.

4. **Answer:** (4) This is correct because these compounds have been identified as strong acid, weak acid, neutral substance, weak base and strong base. If you answered choice (5), you probably did not read the question carefully—the correct order is exactly reversed.

5. **Answer:** (3) This is correct because bases neutralize acids. Choice (1) is incorrect because a neutralization reaction produces a salt. Choices (2) and (4) are incorrect because they are properties of acids. Choice (5) is irrelevant.

6. **Answer:** (3) This is correct because both salt and water are neutral substances. Choices (1) and (2) are incorrect because they are pH values of acids, and choices (4) and (5) are incorrect because they are pH values of bases.

GED Mini-Test pp. 123–124

1. **Answer:** (1) Choice (2) is incorrect because sodium bicarbonate decomposes to form carbon dioxide. Choices (3) and (4) are incorrect because carbon dioxide is the product of the reaction that makes dough rise. Choice (5) is incorrect because sodium bicarbonate is one of the substances initially present in baking powder.

2. **Answer:** (5) This is correct because it is logical that a gas produced would cause the dough to rise. Choices (1), (3) and (4) are incorrect because they are not gases, and choices (1) and (4) are not produced when sodium bicarbonate reacts. Choice (2) is a gas, but there is no mention of oxygen in the passage.

3. **Answer:** (4) This is correct because the passage states that sodium bicarbonate reacts with acids to form carbon dioxide and water. The other choices are incorrect because no other conditions are cited that would cause carbon dioxide and water to be produced from sodium bicarbonate.

4. **Answer:** (3) This is correct because tartrate plus water produces acid, which in turn reacts with the sodium bicarbonate in baking powder to liberate carbon dioxide gas. Choice (1) is incorrect because heat causes sodium bicarbonate to decompose. Choices (2), (4) and (5) are incorrect because no mention is made of reactions involving oxygen.

5. **Answer:** (2) Choices (1) and (4) are incorrect because neither of these substances contains sodium. Choices (3) and (5) are incorrect because they do not contain sodium and are by-products of the reaction.

6. **Answer:** (3) This is correct because ammonia is produced as a by-product in step 3, then recycled to be used again as a reactant in step 2. The other choices are incorrect because they do not reappear.

ANSWERS AND EXPLANATIONS

REVIEW Chemistry

In this section you have learned about matter and the different states of matter. You have learned about an atom, which is the smallest particle of an element, and about a molecule, which is the smallest particle of a compound. You have learned how changes in matter are produced by chemical reactions, and how energy and entropy changes accompany chemical reactions. You have also learned about special groups of compounds known as acids, bases and salts.

In the material that follows you will learn more about the invisible world of molecules.

DIRECTIONS: Choose the one best answer for each item below.

Items 1–4 refer to the following passage.

Many of the properties of matter can be explained in terms of molecular motion. The theory that summarizes these explanations is called the kinetic molecular theory. All molecules are constantly in motion. The type and extent of this motion depends upon whether the molecules are in a solid, liquid or gas.

The molecules of a gas exhibit constant, random motion. For all practical purposes, there are no forces of attraction between gas molecules. You can understand that this is true if you think of how quickly a gas will escape from a container and "disappear."

The molecules of a liquid do not move nearly as freely as the molecules of a gas, nor do they have as much energy. Forces of attraction between liquid molecules keep them bound together. That is why, when you pour a liquid, it stays together and does not spread apart the way a gas does.

Molecules in a solid have the least freedom of motion. The only movement allowed a solid molecule is vibration. Molecules of a solid are held very tightly in place, which is why solids have both definite volume and definite shape.

1. The above passage discusses a relationship between which of the following two factors?

   (1) states of matter and molecular motion
   (2) temperature and molecular energy
   (3) states of matter and sizes of molecules
   (4) sizes of molecules and molecular speed
   (5) temperature and molecular attractions

2. A block of ice changes into liquid and then into steam. What conclusion can be drawn from this information?

   (1) Molecular motion increases as temperature decreases.
   (2) Molecular motion increases as temperature increases.
   (3) Molecular motion becomes more random as temperature decreases.
   (4) Molecules collide more frequently when temperature decreases.
   (5) Phase changes occur more rapidly at low temperatures than high temperatures.

GO ON TO THE NEXT PAGE.

3. Which of the following statements can be inferred from the passage?

   (1) Large molecules move faster than small molecules.
   (2) Molecules in gases have less energy if the gas is in a small container.
   (3) You can feel the molecules vibrating when you hold certain solids.
   (4) Molecules in solids remain in fixed positions, while molecules in liquids and gases are free to move from place to place.
   (5) Molecules of a gas will spread out in a regular pattern as they move to fill a container.

4. In a carbonated beverage you can see bubbles of gas rising to the surface. According to the information provided, these bubbles are caused by which of the following?

   (1) vibrations of gas molecules
   (2) collisions between liquid molecules and gas molecules
   (3) rapidly moving gas molecules moving through less freely moving liquid molecules
   (4) forces of attraction between liquid molecules and gas molecules
   (5) randomly moving liquid molecules pushing against less freely moving gas molecules

Items 5–8 refer to the following diagrams and passage.

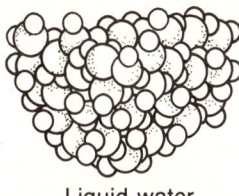
Liquid water

Ice structure

Water is an unusual substance in that its density decreases, rather than increases, as the temperature drops. This unusual property of water is due to the fact that the crystal arrangement of ice incorporates a lot of space between molecules. The density of a substance is equal to the mass of substance per unit volume. Thus, it stands to reason that loosely packed molecules will be less dense than tightly packed molecules.

5. Compared to ice molecules, liquid water molecules are

   (1) more numerous
   (2) larger
   (3) more tightly packed
   (4) different in shape
   (5) softer

6. It can be inferred from the information provided that most substances

   (1) freeze at higher temperatures than water
   (2) expand as temperature increases
   (3) are denser in the liquid state than in the solid state
   (4) are less dense than water
   (5) have molecules that are more tightly packed in the liquid state than in the solid state

7. A concrete swimming pool is cracked when the water inside the pool freezes. Which of the following statements explains why this happens?

   (1) Ice takes up more space than an equal mass of liquid water.
   (2) Water molecules increase in size as the temperature drops.
   (3) The density of ice is greater than the density of liquid water.
   (4) Molecules in concrete spread out when the temperature drops.
   (5) Ice crystals have sharp edges that damage the concrete.

8. The surface of a pond is frozen, but the water underneath is not. Which of the following statements explains why this is so?

   (1) Water molecules are more numerous at the surface of a pond than at the bottom.
   (2) Ice floats on water because it is more dense than water.
   (3) Water is colder near the surface of a pond than at the bottom.
   (4) Water molecules move less at the bottom of a pond than at the surface.
   (5) Ice floats on water because it is less dense than water.

Check your answers to the Review on page 157.

OVERVIEW
Physics

Skater displaying the laws of motion through stop-action photography.

thermodynamics
the study of the relationship between heat and mechanical energy

Physics is the study of matter and energy. It is the branch of science that answers the questions, Why does this happen? and How does this work?

In the first lesson of this section you will learn how a scientific law describes the behavior of matter and energy. You will read in detail about two important sets of laws—the laws of thermodynamics and Newton's laws of motion. The laws of **thermodynamics** explain the relationship between heat and mechanical energy. Because heat is a form of energy, it can be converted into other forms of energy. When heat is converted into mechanical energy, it is able to do work. You will discover how this process takes place in the gasoline engine of an automobile.

In this lesson you will also learn how motion is described by Newton's first and second laws. You will discover that motion is changed by **forces,** and that forces can be represented by mathematical quantities called **vectors.** You will also learn how machines can change the size and direction of a force to make work easier.

In the second lesson you will discover what scientists mean when they say "opposites attract" as you learn about electricity and magnetism. You will discover how electricity is related to the tiny, charged particles inside an atom. You will also find out where the **electric current** in your home "comes from"—and how an **electric circuit** is needed to make the flow of current possible.

In this lesson you will also read about the discovery of the first "magnet" over 2,000 years ago. You will learn about the properties of magnets, and how an important relationship exists between electricity and magnetism.

In the last lesson you will read about the properties of waves. A **wave** is a rhythmic disturbance that transfers energy from one place to another. You will find out how a sound wave transfers energy by disturbing the molecules of the medium through which it travels. You will also learn about light waves, and how they are able to travel through space without a medium.

In this lesson you will discover what happens to a light wave when it strikes a substance and bounces back, or when it passes from one material into another. The behavior of light in these instances is described by the laws of **reflection** and **refraction.**

In the review lesson of this section you will be introduced to nuclear physics. You will learn what happens to **radioactive** elements, and how the splitting of atomic nuclei can produce huge quantities of useful energy.

force
a push or pull acting on an object

vector
a quantity that has both magnitude and direction

electric current
the flow of electrons through a wire

electric circuit
a continuous, unbroken pathway over which electric current can flow

wave
a rhythmic disturbance that travels through space or matter

reflection
the bouncing back of light from a surface

refraction
the bending of light as it passes from one medium into another

radioactivity
the release of energy and matter from an unstable atomic nucleus

15 Motion, Thermodynamics

A law in science is a "rule" that describes the behavior of matter and energy. Unlike a civil law, a **scientific law** does not dictate or change behavior; it simply describes what many people have observed over a period of time. If further observations contradict a scientific law, then the law is changed.

Two important sets of laws of physics are Newton's laws of motion and the laws of thermodynamics. In this introduction you will read about the laws of thermodynamics. In the Practice part of this lesson you will learn about the laws of motion.

Thermodynamics is the study of the relationships between heat and mechanical energy. Because heat itself is a form of energy, it can be converted into other forms of energy—and other forms of energy can be converted into heat. You can experience this conversion if you rub your hands briskly together several times. Your hands will begin to feel warm because the mechanical energy of rubbing them together is being changed into heat.

The first law of thermodynamics states that all energy is conserved. This means that no energy will be lost in the process of converting one form of energy into another.

The second law of thermodynamics describes the way heat moves from one object to another. If you put your hands around a bowl of hot soup, your hands get very warm. If you hold the soup long enough, the bowl will begin to get cooler as it loses heat to your hands. This happens because heat always flows from hotter objects to colder ones.

The transfer of heat from one object to another can be explained in terms of molecular motion. The molecules in hot objects have more energy than the molecules in cold objects. For example, a hot piece of iron is plunged into water at room temperature. As the two substances come in contact with each other, the higher energy iron molecules begin to collide with the water molecules, transferring some of their energy in the process. As a result the iron loses some of its heat and the water gains heat. If heat is a form of energy, you may wonder what "coldness" is. Actually, there is no such thing as coldness—coldness is simply the absence of heat. As energy in the form of heat leaves an object, the object becomes cold.

130 INTRODUCTION

Strategies for READING

Distinguish Conclusions from Supporting Statements

This skill involves identifying a correct statement that would justify a conclusion. This skill may involve identifying a main idea and supporting details; fact and opinion; cause and effect.

Very often in science you must explain why certain things have happened, or predict what will happen as a result of certain conditions. In order to do this you must be able to distinguish a **conclusion** from its **supporting statements** and identify those statements that will justify your explanations or predictions.

In the passage on the previous page, the author discusses two laws that describe the behavior of heat (supporting statements). Suppose that you want to predict what will happen if a piece of hot apple pie comes in contact with a scoop of cold ice cream. Based on the second law of thermodynamics, you can safely say that the apple pie will lose some heat and the ice cream will gain some heat (conclusion).

Example

DIRECTIONS: Use the information on this and the preceding page to choose the one best answer for the item below.

1. As you hold an ice cube in your hand, your hand begins to feel cold and the ice cube begins to melt. Which of the following statements is a conclusion?

   (1) Coldness from the ice flows onto your hand.
   (2) Heat leaves your hand and is absorbed by the ice.
   (3) Molecules of ice lose energy and flow onto your hand as water.
   (4) Molecules in your hand gain energy as a result of holding the ice.
   (5) Energy changes from one form to another as it goes from ice to your hand.

Answer: (2) Choice (1) is incorrect because there is no such thing as "coldness." Choices (3) and (4) are incorrect because the heat transfer from your hand to the ice causes molecules in your hand to lose energy and molecules in the ice to gain energy. Choice (5) is incorrect because it takes out of context an idea that is irrelevant to the situation.

READING: ANALYSIS

Practice

HINT ▷ Plan your study time realistically—in small chunks. (Plan ahead!) Take notes and use your notes to study. Map out the main idea and supporting details to help you understand what you are reading.

DIRECTIONS: Choose the one best answer for each item below.

Items 1–4 refer to the following passage.

Motion is changed by forces. It takes force to start motion, and it takes force to stop motion. It also takes force to change the direction of motion.

Newton's first law of motion states that an object at rest tends to remain at rest, and an object in motion tends to keep moving in a straight line at the same velocity, unless acted on by an unbalanced force. The tendency of an object to keep moving or remain at rest is called **inertia.** You can understand how inertia works if you picture yourself standing in a bus that is stopped. As the bus starts up and moves forward, your body will seem to move backward. What is actually happening is that your body is maintaining its rest position while the bus is pulling forward away from it.

Newton's second law states that an object will accelerate in the direction of the force that acts on it. The size of the force will be directly proportional to the amount of acceleration and the mass of the object. This law is summed up by the formula, F = ma (force = mass × acceleration).

1. Force is measured in units called newtons. If a force of 8 newtons is applied to a 2-kilogram ball, and a force of 6 newtons is applied to a 3-kilogram ball, how will the accelerations of the balls compare?

   (1) Both accelerations will be the same.
   (2) The acceleration of the second ball will be twice that of the first.
   (3) The acceleration of the first ball will be twice that of the second.
   (4) The acceleration of the first ball will be four times that of the second.
   (5) The acceleration of the second ball will be four times that of the first.

2. According to Newton's laws, an unbalanced force would be required to make which of the following situations occur?

   (1) A cyclist traveling at 15 m.p.h. continues to travel at the same speed in the same direction.
   (2) A person who stood through the first hour of a sold-out lecture stands through the second hour.
   (3) A passenger on the subway sits reading while the train travels four miles in four minutes.
   (4) Two children forehead-to-forehead stare into each other's eyes.
   (5) A car traveling at 40 m.p.h. goes around a curve at the same speed.

GO ON TO THE NEXT PAGE.

3. Which of the following situations is an example of Newton's first law?

   (1) A man finds that pushing a full wheelbarrow takes more force than pushing an empty wheelbarrow.
   (2) A passenger in a car going 60 m.p.h. slams into the windshield when the car stops suddenly.
   (3) Passengers notice that a bus rides more smoothly at 50 m.p.h. than at 25 m.p.h.
   (4) A car stuck on an ice patch is able to move when a blanket is placed under the back wheels.
   (5) A football player intercepts a pass and runs in the opposite direction from which the ball was thrown.

4. Football teams use large players in the defense lines and smaller players in the backfield to run and catch passes. What is an advantage of this strategy?

   (1) Small players accelerate quickly, while large players apply force to stop the motion of opponents.
   (2) Large players remain at rest, while light players remain in motion.
   (3) Light players can catch passes, while large players can tackle.
   (4) Large players tend to move in a straight line, while light players can change direction easily.
   (5) The force needed to stop a large player is much greater than the force needed to stop a small player.

Items 5–6 refer to the following passage and diagrams.

Diagram 1
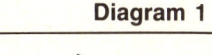

Diagram 2
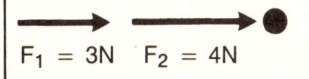

Diagram 3
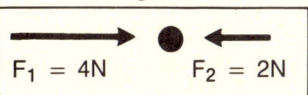

The motion of an object is affected by a force's magnitude and direction, which can be indicated by vectors (represented by arrows). The combined effect of forces acting on an object is called the **resultant force**.

If two forces act in the same direction on an object, the resultant is obtained by adding the forces. In Diagram 1 the resultant is seven newtons to the right.

If two forces act on an object from exactly opposite directions, the magnitude of the smaller is subtracted from the magnitude of the larger — and the resultant will act in the direction of the larger. In Diagram 2, the resultant force is two newtons to the right.

Two forces will cancel each other out if applied in the same magnitude in opposite directions. In the case of Diagram 3, the resultant force is zero.

5. Four cars are in a chain collision. The force experienced by the second car is

   (1) the same as the force experienced by all the other cars
   (2) greater than the force experienced by all the other cars
   (3) less than the force experienced by all the other cars
   (4) greater than the force experienced by the first car but less than the force experienced by the other cars
   (5) less than the force experienced by the first car but greater than the force experienced by the other cars

6. If the ball shown in Diagram 2 were traveling at a speed of eight meters per second at the time forces F1 and F2 were applied, what would happen to the ball's motion?

   (1) It would continue to travel to the right at the same speed.
   (2) It would reverse directions and travel to the left at the same speed.
   (3) It would come to a stop.
   (4) It would continue to travel to the right but at a speed less than eight meters per second.
   (5) It would continue to travel to the right but at a speed greater than eight meters per second.

GED Mini-Test 15

TIP: Be careful about changing an answer. Do not change it *unless* you have found new information that led you to another choice. Most often your first choice is the best.

DIRECTIONS: Choose the one best answer for each item below.

Items 1–4 refer to the following passage.

Energy is the capacity to do work. Because heat is a form of energy, it must also have the capacity to do work.

Before examining how heat is able to do work, you must first understand what work is. You may think of work as earning a living or doing chores. But a scientist defines work as force acting on an object, causing it to move. You do work when you drive a nail with a hammer, because the force you are exerting is causing the nail to move.

One of the most familiar ways in which heat is used to do work is in a heat engine. The gasoline engine that powers a car is an example of a heat engine.

In a gasoline engine, the burning of fuel produces hot gases that move into cylinders containing pistons. When an electrical spark ignites the gases, they explode and push against the pistons. The movement of the pistons eventually transfers energy to the wheels of a car through a series of shafts and gears. The wheels exert a force against the ground to move the car.

Most of the energy produced by gasoline engines is wasted. Only about 12% of the energy provided by the fuel is used to power the car. As a result, much of the heat produced by a gasoline engine is released through the exhaust pipes into the atmosphere.

1. According to the information provided, the *initial* force that sets a car in motion is which of the following?

   (1) the combustion of gasoline
   (2) the push of hot gases on a piston
   (3) the push of hot gases on shafts and gears
   (4) the push of pistons on shafts and gears
   (5) the push of the wheels against the ground

2. According to the information provided, in which of the following situations would you be doing work?

   (1) studying for a physics test
   (2) standing in line for an hour
   (3) holding a heavy bag of groceries for an hour
   (4) hitting a tennis ball across the net
   (5) dropping a ball out of a second-storey window

GO ON TO THE NEXT PAGE.

3. In a gasoline engine, hot gases release energy to power a car by

   (1) releasing oxygen
   (2) exerting pressure
   (3) transferring heat to other substances
   (4) releasing heat to the atmosphere
   (5) changing water into steam

4. Within the same town, the temperature on heavily traveled streets tends to be higher than the temperature on streets with little traffic. Why is this so?

   (1) The pavement on heavily traveled streets absorbs more sunlight than the pavement on less-traveled streets.
   (2) Heat released from many cars raises the temperature of heavily traveled streets.
   (3) The wearing away of road surfaces raises the temperature of heavily traveled streets.
   (4) Heavily traveled streets have fewer trees to provide shade.
   (5) There are more heat-generating industries located near heavily traveled streets than near less-traveled streets.

Items 5–6 refer to the following passage.

A machine is a device that makes work easier. Sometimes it is important to know the actual advantage of using a machine. If a man uses a pulley system to raise 200 pounds (resistance) and applies 25 pounds of force (effort) to do so, we know the pulley system multiplied his effort by eight (the mechanical advantage).

An inclined plane can also be used to raise objects. Instead of lifting an object straight up, a person uses a slanted surface. Less force is required, but the object must be moved a greater distance. Its mechanical advantage is found by dividing the length of the slanted surface by the height of the plane.

5. A woman wishes to raise a box four meters off the ground. Which of the following inclined planes would give her the *greatest* mechanical advantage?

   (1) an inclined plane three meters high and three meters long
   (2) an inclined plane four meters high and three meters long
   (3) an inclined plane four meters high and four meters long
   (4) an inclined plane four meters high and six meters long
   (5) an inclined plane three meters high and four meters long

6. The pulley with the *greatest* mechanical advantage is the one that requires

   (1) 40 pounds of effort to raise a 480-pound rock
   (2) 10 pounds of effort to raise a resistance of 50 pounds
   (3) 80 pounds of effort to raise an 800-pound boulder
   (4) 15 pounds of effort to raise a 300-pound box
   (5) 50 pounds of effort to raise a resistance of 650 pounds

Check your answers to the GED Mini-Test on page 136.

Answers and Explanations

Practice pp. 132–133

1. **Answer:** (3) According to the equation $F = ma$, the acceleration of the first ball will be 4 meters per second^2 (the way acceleration is expressed) and the second ball's will be 2 meters per second^2. The remaining choices indicate a misuse of the formula.

2. **Answer:** (5) This is correct because force is required to change the direction of motion. All the other choices are incorrect because they describe either objects at rest or objects continuing in motion in a straight line.

3. **Answer:** (2) This is correct because the passenger's body keeps moving forward even though the car stops. Choice (1) is incorrect because it is an example of Newton's second law. Choices (3), (4) and (5) are incorrect because they describe types of motion that are not related to the information provided.

4. **Answer:** (1) This is based on Newton's second law. Choice (5), while a valid statement according to Newton's second law, does not really answer the question (see this lesson's Tip). Choice (2) is a distortion of Newton's first law. Choices (3) and (4) may be true, but they are totally irrelevant.

5. **Answer:** (5) This is correct because, according to Situation A, forces acting on an object in the same direction are added. Thus the first car would feel the impact of all the other cars, while the second car would feel the impact of all but the first car. The other choices contradict the correct answer.

6. **Answer:** (5) This is correct because the net force acting on the ball would accelerate it to the right. All the other choices contradict the correct response.

GED Mini-Test pp. 134–135

1. **Answer:** (2) Choice (1) is incorrect because combustion is not a force, although the heat of combustion can produce a force. Choice (3) is incorrect because hot gases do not act directly on shafts and gears. Choices (4) and (5) are incorrect because they are forces that occur later in the process of moving the car.

2. **Answer:** (4) This is correct because it is the only situation in which you are producing a force that is acting on an object and causing it to move. In choice (5), which is tricky, *you* are not producing the force that is moving the ball—you are just allowing gravity to pull it to earth!

3. **Answer:** (2) This is correct because the gases push against the pistons as a result of exploding upon ignition. Choice (1) is incorrect because igniting the gases would require oxygen rather than release it. Choice (3) is incorrect, because, although heat is certainly transferred to surrounding substances, this is not what powers the car. Choice (4) is incorrect because this is how heat energy not used to power the car is wasted. Choice (5) is irrelevant.

4. **Answer:** (2) All of the other choices could be true—although choice (1) sounds a little far-fetched—but they do not relate to the information provided.

5. **Answer:** (4) This is correct because the mechanical advantage is found by dividing the length by the height. Choices (2) and (3) give a smaller mechanical advantage. Choices (1) and (5) do not provide the necessary height to lift the box four meters.

6. **Answer:** (4) This is correct because the mechanical advantage is 300/15, or 20. The mechanical advantages of the other choices are 12, 5, 10 and 13.

16 Electricity and Magnetism

Have you ever heard the cliché "opposites attract"? People usually use this phrase to describe opposite personality types. But this phrase can also describe the forces that exist in two important areas of physics—electricity and magnetism. In this lesson you will read about the fascinating world of magnetism.

You may think of electricity as turning on the lights or starting up an electric motor. Certainly these are important uses of electricity—but what exactly *is* electricity? Where does it come from? And how does it power the appliances that you use every day?

In order to answer these questions you must first go back to the structure of the atom that you studied in Lesson 13. You will recall that atoms are made up of protons, electrons and neutrons. Protons and electrons have a basic property called **electric charge.** Protons are positively charged and electrons are negatively charged. Neutrons have no charge—as their name implies, they are neutral.

If protons and electrons did not have opposite charges, an atom would simply fly apart. It is the force of attraction between the positively charged protons and the negatively charged electrons that holds an atom together. The structure of an atom illustrates the basic rule of electric charge: Unlike charges attract each other while like charges repel each other. You can remember this rule easily if you just think of the old saying "opposites attract."

If one charged particle comes in contact with another charged particle, it will experience a force. The **force** will be one of attraction if the particles are of unlike charge and repulsion if the particles are of like charge. The area of force that surrounds a charged particle is called an **electric field.** The strength of an electric field varies according to the distance from the charged particle—as the distance increases, the strength of the field decreases.

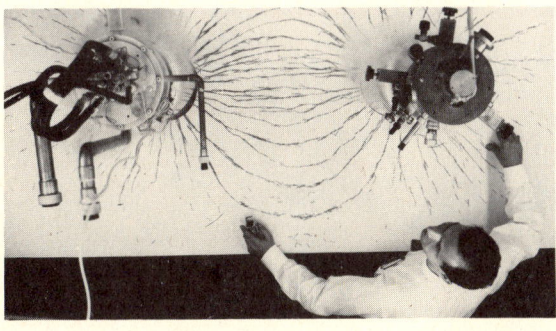

Extremely high magnetic fields at the General Electric Research Laboratory.

INTRODUCTION 137

Strategies for READING

Identify Cause and Effect Relationships

A cause and effect relationship indicates how one thing affects another. A cause is what makes something happen. An effect is what happens as a result.

When you flip on a light switch, you cause the lights to come on. This is a simple example of cause and effect—an obvious action, or **cause** (what happened), produces an obvious result, or **effect**. Sometimes in science, however, cause and effect relationships are not so obvious. Sometimes it is an underlying property or principle that produces a given result, rather than a specific action.

In the passage on the previous page, you read about the electrically charged particles that exist within an atom. It is the attraction between these positively and negatively charged particles that causes an atom to hold together. These attractive forces, then, are the cause—and a stable atom is the effect, or result.

Example

DIRECTIONS: Use the information on this and the preceding page to choose the one best answer for the item below.

1. A negatively charged strip of metal is suspended from the ceiling by a string. When another strip of metal is brought close to it, the first strip bends away. Which statement explains why this effect happens?

    (1) The second strip is positively charged.
    (2) The second strip is negatively charged.
    (3) The second strip contains fewer electrons than the first strip.
    (4) The second strip has a larger electric field than the first strip.
    (5) The atoms in the second strip repel the atoms in the first strip.

Answer: (2)   This is correct because like charges repel each other. Choice (5) is incorrect because it is positively and negatively charged particles, not atoms, that attract and repel each other. Choices (3) and (4) are irrelevant.

READING: ANALYSIS

Practice

HINT Pay attention to key words—those that define: that is, means, such as, is called—those that show contrast: however, yet, still, but, instead of—those that indicate cause: because (of), since, due to—and those that signal an effect: as a consequence, as a result, thus, therefore and so. They can help you understand what you are reading on a GED test.

DIRECTIONS: Choose the <u>one</u> best answer for each item below.

Items 1–4 refer to the following passage.

Have you ever rubbed a balloon against your sleeve and stuck it to the wall? You were able to do this because the balloon became electrically charged.

Electrons in atoms are free to move. As you rub a balloon on your sleeve, electrons from the cloth move onto the balloon. This excess of electrons gives the balloon a negative charge. When the balloon comes near the wall, it repels the electrons in the wall, leaving the wall positively charged. Thus, the negatively charged balloon and the positively charged wall attract each other.

The ability of electrons to move from one place to another makes possible electric current. Electric current is the flow of electrons through a wire, and it is what powers your appliances.

Electrons flow through a wire in much the same way as water flows through a hose. Just as water pressure pushes water through a hose, a source of energy pushes electrons through a wire. This source of energy is called **voltage.** Voltage is measured in units called **volts.** A battery marked "9V" means that the battery supplies nine volts of energy to move electrons.

1. In the above passage, water pressure is compared to voltage because

   (1) both provide energy to move electrons
   (2) both make possible the flow of a substance from one place to another
   (3) both are measured in units called volts
   (4) both involve the transfer of electrically charged particles
   (5) both can be used to power a battery

2. Excess electrons on an electrically charged object eventually leave the object, returning it to its neutral state. A person who wishes to stick balloons to the wall as party decorations should

   (1) place the balloons on the wall just before the party begins
   (2) place the balloons far apart
   (3) place the balloons on the wall at least a day ahead of time
   (4) rub the balloons on several different types of cloth
   (5) place the balloons as close to the ceiling as possible

GO ON TO THE NEXT PAGE.

3. If you scuff your feet across a wool carpet, you may experience a "shock" when you touch a metal object. Which statement explains why this effect happens?

(1) Electric charge builds up on the rug, then is transferred to the metal object.
(2) Electrons from the carpet move onto your feet, causing you to temporarily acquire an electric charge.
(3) Electrons flow from your feet to the carpet, causing the carpet to acquire an electric charge.
(4) Electrons in the metal object are repelled by electrons on the carpet.
(5) Scuffing across the carpet causes an increase in voltage, which then produces a shock.

4. Which of the following statements could *not* be supported by the information provided?

(1) Protons flowing onto an object cause the object to become positively charged.
(2) Low voltage could cause a power "brownout" in a town or city.
(3) If two balloons that had been rubbed on cloth were brought close together, they would probably repel each other.
(4) A torn wire would cause a loss of electricity just as a leaky hose would cause a loss of water.
(5) Current flow increases as voltage increases.

Items 5–6 refer to the following passage and diagram.

In order for electric current to flow, electrons must have a closed, continuous pathway over which to travel. Such a pathway is provided by an electric circuit. An **electric circuit** consists of a source of electrons, a load or resistance and a switch. In the diagram the source of electrons is an eight-volt battery. The wavy lines you see are resistors. A resistance can be a light bulb, an appliance or a motor—anything that uses electrical energy.

Electrical current (I) is measured in amperes ("amps" for short). Resistance (R) is measured in ohms. Current, resistance and voltage are related by the equation V = I × R.

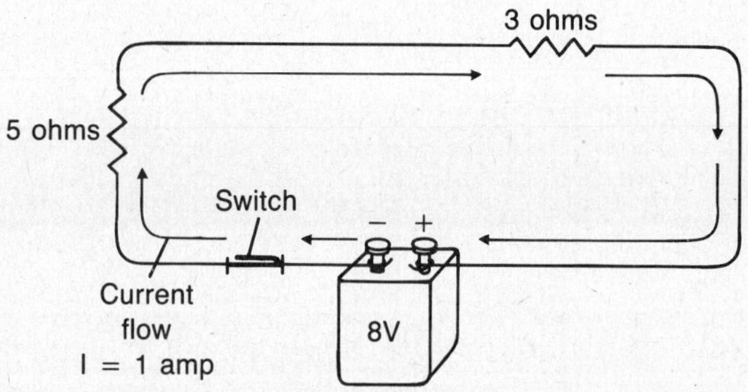

Electric Circuit

5. If the voltage in the circuit above were reduced to four volts, what would be the effect on the current?

(1) It would stay the same.
(2) It would be cut in half.
(3) It would double.
(4) It would increase four times.
(5) It would be equal to the voltage.

6. An "overload" circuit would probably result in which of the following?

(1) an increase in voltage
(2) a decrease in current
(3) an increase in current flow
(4) a decrease in voltage
(5) a greater number of electrons passing through each resistor

Before you take the GED Mini-Test, check your answers on pages 142–143.

GED Mini-Test 16

TIP When selecting an answer, do not base your choice on its number. If, for example, choice (1) has not been the correct answer for the last three questions, do not be fooled into thinking it is the correct choice for the question you are presently answering.

DIRECTIONS: Choose the one best answer for each item below.

Items 1–4 refer to the following passage.

An important relationship between electricity and magnetism was discovered by Danish scientist Hans Christian Oersted in 1820. Oersted discovered that a magnetic field is formed around a wire conducting an electric current. This phenomenon is known as electromagnetism.

Because of electromagnetism, powerful temporary magnets called electromagnets can be made by wrapping coils of wire around soft iron and passing an electric current through the wire. When the current passing through the wire is turned off, the magnet loses its magnetic properties. When the current is turned back on, the magnet regains its magnetic properties. The strength of the electromagnet will depend upon the number of loops of wire and the size of the current.

1. A piece of crane-like heavy machinery used at construction sites is equipped with a large electromagnet. The electromagnet will make the machine *most* useful for which of the following tasks?

   (1) transporting pieces of iron to distant locations
   (2) picking up pieces of scrap metal on the site and depositing them elsewhere on the site
   (3) generating electricity for the area surrounding the site
   (4) making magnets out of pieces of scrap metal found on the site
   (5) lifting objects too heavy for other types of machines on the site

2. The needle of a compass placed near the wire of an electric circuit is deflected 90° when the circuit is turned on. Which of the following statements explains why this happens?

   (1) The needle has lost its ability to point north.
   (2) The wire in the circuit must be pointing north.
   (3) The needle is responding to the magnetic field produced by the current in the wire.
   (4) Electrons in the needle are repelled by electrons in the wire.
   (5) The compass needle has become an electromagnet.

3. Based on the passage, the relationship between electricity and magnetism is *best* expressed by which of the following statements?

   (1) Electromagnets form magnetic fields.
   (2) Magnets cause electricity.
   (3) Wrapping wire around iron causes magnetism.
   (4) An electric current produces a magnetic field.
   (5) An electric field is produced by an iron magnet.

4. Which of the following hypotheses can be supported by the information provided?

   (1) Magnets make electric current strong.
   (2) Electromagnets are not as strong as natural magnets.
   (3) A magnet can reverse the direction of an electric current.
   (4) The strength of an electromagnet depends upon the material used to make the magnet.
   (5) Magnetism is related to the movement of electrons.

GO ON TO THE NEXT PAGE.

Items 5–6 refer to the following passage.

Over 2,000 years ago, a mysterious stone was discovered that could attract bits of material containing iron. The Greeks who found this stone named it "magnetite"; they had discovered an interesting property of certain materials that we now call magnetism.

Magnetite is an example of a natural magnet. Most of the magnets you have probably used are artificial magnets. The simplest artificial magnet is an iron bar magnet.

If you suspend a bar magnet horizontally on a string and allow it to swing freely, one end of the magnet will always point north. This end of the magnet is called the north magnetic pole. The other end of the magnet, which points south, is called the south magnetic pole.

The area in which magnetic forces can act is called a magnetic field. The magnetic field of a bar magnet is strongest around the poles. If a north pole of one magnet and a south pole of another magnet are brought together, they will attract each other. If two north poles or two south poles are brought together, they will repel each other. Thus the rule for magnetic poles is: *Unlike poles attract each other while like poles repel each other.* Once again, "opposites attract."

5. The discovery of magnetite was important because it

   (1) made artificial magnets possible
   (2) was the first natural iron magnet
   (3) gave scientists information about the properties of iron
   (4) illustrated a property of certain materials known as magnetism
   (5) explained why one pole of a magnet always points north

6. The needle of a compass always points to the north. Which of the following *must* be true about compass needles?

   (1) Compass needles are made of magnetite.
   (2) Compass needles are made of iron.
   (3) Compass needles are magnetic.
   (4) A compass needle has a north magnetic pole but no south magnetic pole.
   (5) Two compass needles will repel each other.

Check your answers to the GED Mini-Test on page 143.

Answers and Explanations

Practice *pp. 139–140*

1. **Answer:** (2) Choices (1), (3) and (4) are incorrect because they apply only to voltage. Choice (5) is incorrect because voltage is supplied by a battery—and water pressure has nothing to do with powering a battery.

2. **Answer:** (1) This is correct because this would minimize the possibility of the balloons losing their negative charge and coming loose from the wall. Choice (3) would work exactly against this idea. Choices (2), (4) and (5) are irrelevant to the information given.

3. **Answer: (2)** This is correct because the situation is similar to the rubbing of a balloon on a sleeve. Choices (1) and (3) are incorrect because you receive the shock, not the carpet—so *you* must have an electric charge as a result of acquired electrons. Choice (4) leaves you out of the action entirely, and choice (5) takes the concept of voltage out of context to create a senseless answer.

5. **Answer: (2)** This is correct because the formula $V = I \times R$ shows that current is directly proportional to voltage as long as resistance stays the same. The other choices contradict the correct answer.

4. **Answer: (1)** This is correct because no mention is made in the passage of protons being able to move (they cannot). Choices (2) and (5) are supported by the passage because electric current is pushed through wires by voltage, and a decrease in voltage would cause a decrease in current. Choice (3) could be supported because both balloons would have a negative charge, and like charges repel each other. Choice (4) is supported by the analogy between the flow of electricity and the flow of water.

6. **Answer: (2)** This is correct because as resistance increases, current decreases. Choices (1) and (4) are incorrect (this is tricky) because a battery made to deliver eight volts can only deliver eight volts. Choices (3) and (5) both refer to an increase in current, and thus contradict the correct answer.

GED Mini-Test pp. 141-142

1. **Answer: (2)** This is correct because the electromagnet will gain and lose its magnetism as the current is turned on and off—thus it is ideal for picking up metal objects, then dropping them again. Choice (1) is incorrect because there is no way of knowing whether this machine is equipped to travel distances. Choice (5) is possible, but not really verifiable since nothing is known about the relative strengths of this and other machines on the site. Choices (3) and (4) are irrelevant.

3. **Answer: (4)** Choice (1) is essentially true, but it does not answer the question because it does not say anything specific about electricity. Choices (3) and (5) are scientifically valid, but this passage discusses only the production of a magnetic field from electric current, not the reverse. Choice (3) is not only too narrow a choice, it also fails to specify that the wire must be conducting an electric current.

5. **Answer: (4)** Choice (3) is plausible, but the discovery of magnetite had much broader implications than just revealing the properties of one metal. Choice (2) is incorrect because magnetite attracts iron—it is not made of iron. Choices (1) and (5) bear no relationship to the information given.

2. **Answer: (3)** Choice (1) is incorrect because it does not adequately relate cause and effect. Also, once the current is turned off, the needle will resume its northward orientation. Choice (5) is incorrect because the magnetism of the compass needle was not produced by the current in the wire. Choices (2) and (4) are irrelevant.

4. **Answer: (5)** The correct choice is (5), because a flow of electrons (electric current) produces a magnetic field. Choices (2) and (4) are not really discussed in the passage. There is no basis at all in the passage for choices (1) and (3).

6. **Answer: (3)** Choices (1) and (2) are possible, but not necessarily true—thus they are not the best choices. Choice (4) is incorrect because any magnetic object must have a north and south pole. Choice (5) is incorrect because it would depend upon how the compass needles were lined up—if the north pole of one needle were near the south pole of the other, they would attract each other.

ANSWERS AND EXPLANATIONS

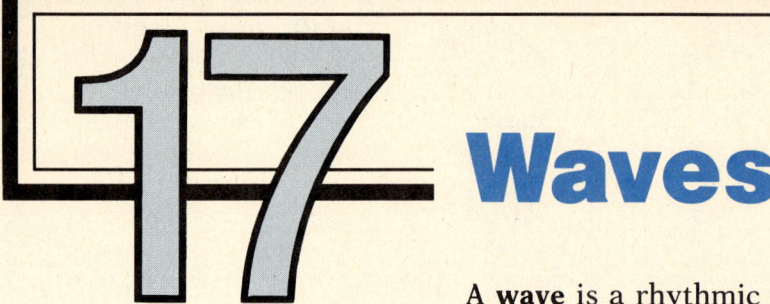

Waves

A **wave** is a rhythmic disturbance that travels through space or matter. Like anything in motion, waves have energy. Waves transfer energy from one place to another. Some familiar types of waves include sound waves, light waves, radio waves, microwaves and ocean waves.

All waves, no matter what their type, have certain basic characteristics. These include amplitude, shape, wavelength and frequency. You can understand these properties most easily if you think of how they apply to an ocean wave.

The **amplitude** of a wave is how high it rises from its rest position. An ocean wave is at rest when the sea is absolutely calm. As the wave rises, it reaches a maximum height and then falls back down again. This high point is called the **crest.** The distance from the rest position to the crest is the amplitude. The greater the amplitude, the greater the energy of the wave.

The **shape** of a wave is created as the wave moves from its rest position to a high point, or crest, then back through the rest position to a low point. The low point of a wave is called the **trough,** and it is the same distance from the rest position as is the crest.

If you have ever watched ocean waves, you have probably noticed that sometimes the waves seem to come very close together, while at other times they seem very far apart. What you were noticing was a difference in wavelength. **Wavelength** is the distance between the crests of two consecutive waves. Waves that appear very close together have a shorter wavelength, while those that appear far apart have a longer wavelength.

Frequency is the number of waves that pass by a given point per unit of time. For example, if you watched an object in the ocean bob up and down ten times in one minute, the frequency of the wave would be ten cycles per minute. In order to be counted as one complete cycle, a wave must pass through both its crest and its trough.

If you know the wavelength and frequency of a wave, you can find its velocity. **Velocity** represents the distance the wave covers per unit of time. If the frequency of the wave is measured in Hertz, and the wavelength is measured in meters, then the velocity in meters per second is given by this equation:

velocity = wavelength × frequency.

Strategies for READING

Assess Adequacy of Data to Support Conclusions

This skill involves applying information in charts, graphs or a written passage to prove or disprove statements.

The scientific process constantly evaluates **data** from experiments, observations or written sources such as reference books, newspaper articles and scientific journals. By carefully evaluating data, scientists are able to draw **conclusions** about the physical world in which we live.

Two things that can help you evaluate data as you read are rules, such as equations and generalizations. In the passage on the previous page, a rule for calculating the velocity of a wave is given, as are general properties of waves. In the example below, assess the adequacy of this information to draw conclusions.

Example

DIRECTIONS: Use the information on this and the preceding page to choose the <u>one</u> best answer for the item below.

1. Which of the following statements can be disproved by the information provided?

   (1) A wave of amplitude 3 meters has a distance of 6 meters from crest to trough.
   (2) A wave with a velocity of 8 meters/second could have a wavelength of 2 meters and a frequency of 4 Hertz.
   (3) A wave with a crest height of 0.3 centimeters has an amplitude of 0.3 centimeters.
   (4) A wave with a frequency of 200 Hertz could have a wavelength of 4 meters and a velocity of 50 meters/second.
   (5) A wave of amplitude 7 meters has more energy than a wave of amplitude 3 meters.

Answer: (4) Because velocity = wavelength × frequency, choice (4) cannot be a true statement. Choices (1) and (3) are supported by the definitions of amplitude, crest and trough. Choice (2) is supported by the equation for velocity. Choice (5) is supported by the general statement that greater amplitude means greater energy.

READING: EVALUATION

Practice

Sometimes finding the best answer to a question requires careful interpretation of the question.

DIRECTIONS: Choose the <u>one</u> best answer for each item below.

Items 1–6 refer to the following passage.

  Sounds are transmitted as waves. In order for a sound wave to travel, it must have a **medium,** a substance capable of transmitting the wave. The medium can be a solid, a liquid or a gas.
  When a sound wave is transmitted, the molecules of the medium vibrate. The wave pushes the molecules back and forth parallel to its line of motion. During one complete cycle of a sound wave, the molecules are pushed together in a **compression,** then allowed to spread out in a **rarefaction.** You can think of the wave's motion as push–pull back, push–pull back. Waves that disturb a medium in this way are called **longitudinal waves**.
  Sound waves travel best through solids, because the molecules are packed tightly together. Elastic solids, such as nickel, steel and iron, transmit sound especially well; inelastic solids, such as sound-proofing materials, transmit sound less well. Liquids are second-best to solids in transmitting sound, and gases are the least effective carriers of sound.
  The speed of sound depends upon the medium. Sound travels through air at room temperature (about 25°C) at 346 meters per second; it travels through air at 0°C at 331 meters per second. The speed of sound in water averages about 1,500 meters per second, while the speed of sound in stone is about 5,970 meters per second.

1. The motion of a longitudinal wave *most* resembles the motion of

    (1) an ocean wave
    (2) a pulsating rope held between two people
    (3) an accordion being played
    (4) a sewing machine needle
    (5) a bicycle on a bumpy road

2. It can be inferred from the passage that a gas is the *least* effective medium for the transmission of sound waves because

    (1) its molecules are too close together
    (2) its molecules are too far apart
    (3) it is too dense
    (4) it is inelastic
    (5) its temperature is too low

GO ON TO THE NEXT PAGE.

3. "Solid state" on audio equipment probably refers to which statement?

   (1) Sound molecules travel fastest in solids.
   (2) Solid components are more durable than other types of components.
   (3) Solids transmit sounds best.
   (4) Solids amplify sounds best.
   (5) Solids are more elastic than gases.

4. The moon has no atmosphere. If you were planning a trip to the moon, which items would be useless to take along?

   (1) a flashlight
   (2) thermal underwear
   (3) oxygen supply
   (4) cassette player with earphones
   (5) tape deck and speaker system

5. On a very cold day a baseball fan notices the ball in the air before hearing the crack of the bat. Which statement offers the *best* explanation?

   (1) The person's ears have become stopped-up by the cold.
   (2) The players are hitting the ball slower because of the cold.
   (3) Some people receive sound waves slower than other people.
   (4) The sound waves have farther to travel than usual.
   (5) Sound waves travel slower in cold air than in warm air.

6. The expression "Put your ear to the ground" probably originated because

   (1) sound that travels through ground will reach your ears before sound that travels through air
   (2) sound waves travel horizontally
   (3) sound waves travel faster at low altitudes than at high altitudes
   (4) some types of sounds travel only through solids
   (5) sound waves are loudest near the ground

Items 7–8 refer to the following passage.

Light waves are **electromagnetic waves.** Unlike sound waves, electromagnetic waves do not need a medium through which to travel. Electromagnetic waves are transverse waves.

Light is the visible part of the electromagnetic spectrum and includes those electromagnetic waves with frequencies between 400 trillion and 750 trillion Hertz. Below these frequencies are invisible waves such as infrared, radio waves, microwaves and radar. Above the frequencies of visible light are other invisible waves, including ultraviolet, X-rays, and gamma rays. All electromagnetic waves travel at the same speed in a given medium. The speed of light in a vacuum is about 300,000 kilometers per second.

7. Compared to radio waves, ultraviolet waves are

   (1) visible, while radio waves are invisible
   (2) lower in energy than radio waves
   (3) lower in frequency than radio waves
   (4) higher in frequency than radio waves
   (5) of longer wavelength than radio waves

8. Which statement can be disproved by the information provided?

   (1) Air is the only medium through which light can travel.
   (2) Light waves can travel through space.
   (3) Gamma rays travel just as fast as microwaves.
   (4) While sound cannot be heard on the moon, light can be seen there.
   (5) No one can see radio waves.

Before you take the GED Mini-Test, check your answers on pages 149–150.

GED Mini-Test 17

TIP Use your reading skills, such as identify the main idea, cause and effect relationships and implications as well as transfer concepts to a new context, assess appropriateness of data and distinguish conclusions from supporting statements, to help you when answering test items. Remember that a key to passing the GED test is reading for understanding.

DIRECTIONS: Choose the one best answer for each item below.

Items 1–2 refer to the following diagrams and passage.

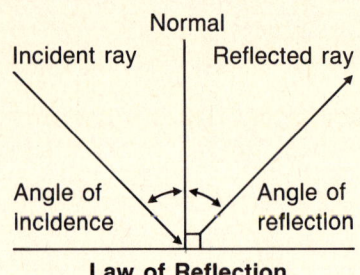
Law of Reflection

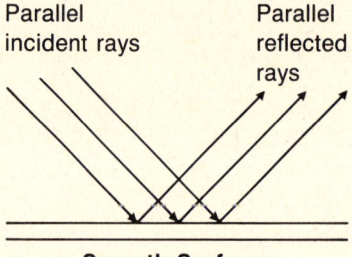
Smooth Surface

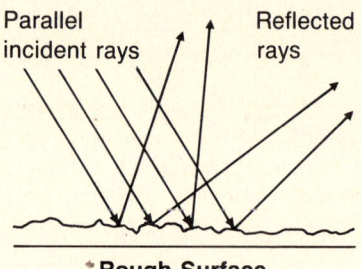
Rough Surface

When a light wave strikes a surface and bounces back, we say that it is reflected. The law of reflection states that the angle of reflection will be equal to the angle of incidence. Both of these angles are measured with respect to the normal (perpendicular).

Because light waves travel in a straight lines, they can be represented by lines that show direction, called rays. A beam of light, such as that produced by a flashlight, contains many parallel rays. If a beam of light strikes a smooth surface such as a mirror, all of the rays will be reflected parallel to one another in the same order in which they originated. If, however, a beam of light strikes a rough surface, each ray will strike the surface at a different angle of incidence, and the light will be scattered in many different directions.

1. A woman wishes to use a mirror to send a signal by flashlight to a man. The angle formed by the woman, the mirror and the man is 60°, so the man will receive the signal at which of the following angles if the woman aims the flashlight toward the mirror?

   (1) 60°
   (2) 15°
   (3) 30°
   (4) 90°
   (5) 120°

2. Dust and other foreign particles in the atmosphere scatter light. An airport control tower located near a city with polluted air would find that its light signals

   (1) appear sharper and clearer than usual
   (2) travel longer distances than usual
   (3) take longer to reach their destinations than usual
   (4) appear fuzzy and diffused
   (5) are all reflected at the same angle

GO ON TO THE NEXT PAGE.

Items 3–4 refer to the following passage.

Light travels at different speeds through different media. For this reason, a light wave may bend as it passes from one medium to another. The bending of light at a boundary between two media is called refraction.

If the speed of light is reduced as it travels from one medium to another, the angle of refraction will be smaller than the angle of incidence. If the speed of light increases as it travels from one medium to another, the angle of refraction will be greater than the angle of incidence. Both angles are measured with respect to the normal.

Light that strikes a boundary between two media perpendicular to the boundary will pass straight through and not be refracted. In order to be refracted, light must strike a boundary at an angle less than 90°.

3. Light travels faster in air than in glass. A light wave striking the side of a glass tank at an angle of 50° will be refracted

   (1) at an angle of 50°
   (2) at an angle greater than 50°
   (3) at an angle less than 50°
   (4) at an angle of 90°
   (5) at an angle greater than 90°

4. A coin can be seen at the bottom of a pool filled with water. When a person tries to retrieve the coin, he discovers that the coin is several feet away from where it appeared to be. Based on the information provided, which of the following statements could explain why this is so?

   (1) Light reflected from the coin is refracted as it crosses the boundary from water to the air.
   (2) The person viewing the coin was looking into the water at an angle of 90°.
   (3) Light waves bouncing back from water interfere with light waves entering it.
   (4) Light entered the water at a 90° angle.
   (5) Light reflected from the coin reaches the air at an angle greater than 90°.

Check your answers to the GED Mini-Test on page 150.

Answers and Explanations

Practice *pp. 146–147*

1. **Answer:** (3)   This is correct because it represents a back-and-forth motion along a straight line. The other choices are incorrect because they represent up-and-down motion.

2. **Answer:** (2)   Choices (1) and (3) are incorrect because substances with molecules close together transmit sound better than substances with molecules far apart. Choice (4) is incorrect because inelasticity was not discussed relative to gases. Choice (5) is incorrect because a gas can be any temperature.

3. **Answer:** (3)   If you answered choice (1), you read the statement too quickly—because there is no such thing as a "sound molecule." Choice (2) is incorrect because it does not relate to the information in the passage. Choice (4) is not discussed in the passage, and the validity of choice (5) cannot be determined from the passage.

4. **Answer:** (5)   This is correct because the sound from the speakers could not travel without air. Choice (4) is incorrect because the earphones and wires attached to the cassette player would serve as a medium to transmit sound to your ears. The other choices are incorrect.

5. **Answer:** (5) Choices (1) and (2), while plausible, do not relate to the subject of the passage. Choices (3) and (4) are not supported or suggested by the information provided.

6. **Answer:** (1) This is correct because sound travels faster through solids than through gases. Choice (2) does not make sense with respect to the question. Choices (3), (4) and (5) are not supported by the information provided.

7. **Answer:** (4) Choice (1) is incorrect because both types of waves are invisible. Choice (2) cannot be determined from the information given. (Actually, ultraviolet waves are higher in energy.) Choice (3) contradicts the correct answer. Choice (5) can be disproved by applying the wave equation given in the introductory passage of this lesson; even without recalling this information, choice (5) would not be the best answer because it cannot be determined directly from the information given.

8. **Answer:** (1) This is correct because electromagnetic waves do not need any medium through which to travel. Choices (2) and (4) are supported by the fact that light can travel without a medium. Choice (3) is valid because all electromagnetic waves travel at the same speed in a given medium. Choice (5) is supported by the fact that radio waves are invisible.

GED Mini-Test   pp. 148–149

1. **Answer:** (3) This is correct because light aimed at an angle of 30° would be reflected at an angle of 30°, totaling the required 60°. Choice (1) is incorrect because the light would be reflected at an angle of 60°, thus hitting a target that would form an angle of 120°. Choice (2) is incorrect because the light would hit a target forming an angle of 30°. The light in choice (4) would be reflected straight back to the sender. Choice (5) is invalid because an incident angle must be between 0° and 90°.

2. **Answer:** (4) Choice (1) is incorrect because the light would be reflected randomly in many directions, resulting in a less focused signal. Choices (2) and (3) are not supported by the information provided. Choice (5) is incorrect because the angle of reflection for each ray will depend upon its angle of incidence.

3. **Answer:** (3) Choices (1) and (2) are incorrect because the angle of refraction will be smaller than the angle of incidence whenever the speed of light is reduced as it passes from one medium into another. Choice (4) would result in the light being refracted along the surface of the glass, something that is physically possible but not discussed in the passage. Choice (5) is impossible because the angle of refraction can never be more than 90°.

4. **Answer:** (1) Choices (2) and (4) are incorrect because they distort the idea that light striking a boundary at 90° is not refracted. Choice (3) is incorrect because it involves reflection and interference, topics not discussed in the passage. Choice (5) is impossible because angles of reflection and refraction must be no greater than 90°.

REVIEW
Physics

In this section you have read about some of the laws that describe the behavior of matter and energy. You have learned about electricity and magnetism. You have also learned about the properties of waves, and how light and sound are transmitted as waves.

In this review section you will learn about nuclear physics. Do not be discouraged if some of this material seems difficult to you. Much of what you will read about has only been recently understood by scientists.

DIRECTIONS: Choose the <u>one</u> best answer for each item below.

Items 1–4 refer to the following passage.

Radioactivity is the release of energy and matter that results from changes in the nucleus of an atom. Only atoms with unstable nuclei are radioactive. Scientists believe that unstable nuclei are caused by an imbalance of protons and neutrons.

All of the elements above and including atomic number 84 (polonium) are radioactive. Many other elements have radioactive isotopes. Isotopes are atoms whose nuclei contain the same number of protons but have a different number of neutrons.

Radioactive elements can change into other elements by a spontaneous process known as radioactive decay. In radioactive decay, the unstable nucleus of a radioactive atom breaks down until the stable nucleus of another element is reached. For example, an atom of uranium will go through thirteen changes until it becomes a stable atom of lead.

What happens to an atom as it changes from one element into another? It emits energy and subatomic particles in the form of radiation. There are three types of radiation.

Alpha radiation consists of two positively charged protons and two neutrons released together in what is known as an alpha particle. Beta radiation consists of tiny, negatively charged beta particles that are actually electrons. Gamma radiation is made up of high-energy electromagnetic waves called gamma rays.

Some radioactive elements undergo alpha decay, while others undergo beta decay. Both alpha and beta decay are nearly always accompanied by the release of gamma rays. Of the three types of radiation, gamma rays are the most harmful. With tremendous penetrating power, gamma rays have the ability to destroy the cells of living things.

GO ON TO THE NEXT PAGE.

1. According to the passage, an atom of a radioactive element does *all* of the following except

    (1) release energy
    (2) increase in size
    (3) change its identity
    (4) release subatomic particles
    (5) emit gamma rays

2. Which of the following would be *most* useful in separating alpha particles from beta particles?

    (1) a microscope
    (2) an electric field
    (3) a powerful lamp
    (4) a magnet
    (5) a block of lead

3. It can be inferred from the passage that human exposure to radioactive elements would be harmful

    (1) only for certain people
    (2) only for certain elements
    (3) only under certain conditions
    (4) only in the presence of beta rays
    (5) in nearly all cases

4. A radioactive isotope of carbon is called carbon-14. An atom of carbon-14 is changed into an atom of nitrogen by radioactive decay. Which of the following statements about these two atoms *must* be true?

    (1) An atom of nitrogen has more energy than an atom of carbon-14.
    (2) An atom of nitrogen is heavier than an atom of carbon-14.
    (3) An atom of nitrogen is less stable than an atom of carbon-14.
    (4) An atom of nitrogen is more stable than an atom of carbon-14.
    (5) An atom of nitrogen contains more alpha particles than an atom of carbon-14.

Items 5–6 refer to the following passage.

   The amount of energy released by an atom during radioactive decay is quite small. Much larger quantities of energy can be released by atoms in a process called nuclear fission. Nuclear fission is the splitting of an atomic nucleus into two smaller nuclei of approximately equal mass. Unlike radioactive decay, a fission reaction must be made to happen. The first nuclear fission reaction was engineered in 1938 by Italian physicist Enrico Fermi.
   The splitting of one atomic nucleus does not release a great deal of energy, but the splitting of many nuclei in rapid succession produces a huge amount of energy. This rapid splitting of many nuclei is called a nuclear chain reaction. It is this process that produces the energy generated at a nuclear power plant. If uncontrolled, a chain reaction results in a nuclear explosion. For this reason, fission reactions take place in a device called a nuclear reactor. It is the purpose of a reactor to control the speed of the reaction and to prevent the escape of radioactive materials into the environment.

5. Which of the following statements about nuclear fission is *not* true?

    (1) Fission reactions occur spontaneously in nature.
    (2) Fission reactions release energy and matter.
    (3) Fission reactions can be controlled.
    (4) Fission reactions involve changes in the nucleus of an atom.
    (5) Fission reactions never took place before the twentieth century.

6. According to the passage, the amount of energy produced by a fission reaction depends upon which factors?

    (1) the type of atoms
    (2) the size of atoms
    (3) the number of atoms and speed of the reaction
    (4) the number of atoms and order of the reaction
    (5) the size of the atoms and speed of the reaction

Answers and Explanations

Life Science Review pp. 70–77

1. **Answer:** (3) According to the diagram and the information presented in the labels, food enters the mouth pore and passes into the gullet. None of the other choices has any relevance to the food particles picked up by the paramecium as it moves through its water environment.

2. **Answer:** (5) According to the information in the labels, the only function of the micronucleus is to control the paramecium's reproduction. Choice (1), which may be confused with the correct choice, controls respiration, protein synthesis and digestion. None of the other choices has anything to do with reproduction.

3. **Answer:** (1) Since the contractile vacuoles remove extra water from the paramecium, it can be assumed that without a contractile vacuole, water would accumulate in the paramecium, causing it to enlarge. Choice (2) states the opposite, and it is not logical. Choice (3) and (5) are irrelevant. Choice (4) is the function of the anal pore.

4. **Answer:** (4) According to the diagram, a food vacuole forms at the end of the gullet. Choice (3) might be confused with the correct answer if the diagram is not studied carefully. Choices (1), (2) and (5) have no relevance to food vacuoles.

5. **Answer:** (1) According to the diagram, the heat from the sun is responsible for all evaporation. Choices (4) and (5) can be thought of as resulting indirectly from evaporation. They are, therefore, not the best answers. Choices (2) and (3) result from chemical reactions taking place in living organisms.

6. **Answer:** (3) According to the diagram, the water falling from clouds is called precipitation. Choice (1) is a chemical reaction that occurs in plants. Choice (2) is the result of water seeping through porous earth. Choice (4) is the effect of the sun's heat on water. Choice (5) is a chemical process that occurs in animals.

7. **Answer:** (5) According to the diagram, water flowing to lakes, rivers and oceans is called ground water. Choices (1) and (4) refer to water entering the atmosphere. Choice (2) refers to water leaving the atmosphere. Choice (3) refers to underground water that does not flow into lakes, rivers or oceans.

8. **Answer:** (3) The evaporation of water given off through transpiration in plants is one source of water in the atmosphere. Choice (4), which may be confused with the correct answer, is the form in which water occurs in the atmosphere and is not a source of atmospheric water. Choices (1) and (2) refer to water returning to the earth. Choice (5) refers to a zone of underground water that is more or less stationary.

9. **Answer:** (1) Since forests are made up of vast numbers of plants, and plants supply water to the atmosphere through transpiration, destroying large amounts of forest land could affect the amount of water in the atmosphere. Choice (2) states the opposite of the correct answer. Choices (3) through (5) are not relevant to the question.

10. **Answer:** (4) According to the diagram, respiration in animals returns some water to the atmosphere. Choices (1) and (5) refer to processes that take place in plants. Choices (2) and (3) refer to non-living, or abiotic, ways of returning water to the atmosphere.

11. **Answer:** (3) According to the chart, mammals first appeared in the Jurassic period. Each of the other choices names a different period. Choices (1) and (2) name periods in which mammals occur, but not for the first time.

12. **Answer:** (5) According to the chart, fish evolved first about 600 million years ago. They were followed by amphibians, reptiles, birds and mammals.

13. **Answer:** (2) Reading the chart carefully, it shows that the first human-like animals appeared about 70 million years ago. Choices (1), (3) and (4) name dates when other animals first appeared. Choice (5) is the age of the earth as shown by the chart.

14. **Answer:** (4) The key phrase in this question is "*most* recently." That means the last plants to evolve, and they were the flowering plants. Each of the other choices names plants that evolved before, or are older than, flowering plants.

15. **Answer:** (2) By studying the chart carefully and comparing the oldest organisms to the more recent, it is clear that life has evolved from the simple to the complex. Choice (1) states the opposite. Choices (3) and (4) are obviously incorrect since the chart covers 4.5 billion years. The chart does not mention catastrophes.

16. **Answer:** (4) Choosing the correct answer depends upon reading the question carefully. The key word is "era." According to the chart, dinosaurs lived during the Mesozoic era. Choice (3) also names a time when dinosaurs lived, but it names a period and is, therefore, incorrect. Choices (1), (2) and (5) name other eras.

17. **Answer:** (1) By studying the diagram, it can be concluded that the tadpole uses its tail for swimming. You could also assume that, since the tadpole's tail makes it look like a fish, it aids the tadpole in swimming. Nothing about the look of the tail would make you conclude that it was used for any of the other choices.

18. **Answer:** (2) In order to arrive at the correct answer, the question must be read carefully. It asks you to compare the adult frog's mouth to the tadpole's mouth. Choice (2) does this accurately. Choice (1) compares the tadpole's mouth to the adult frog's and is, therefore, the wrong answer. Choices (3) through (5) are obviously incorrect.

19. **Answer:** (5) You are probably familiar with the gills of fish, which allow them to breathe underwater. A likely conclusion would be that the gills of tadpoles function in the same way. In addition, the tadpole loses its gills when it becomes a frog and lives on land. Again, a likely conclusion is that the gills allow the tadpole to breathe underwater and are not necessary in a terrestrial environment.

20. **Answer:** (3) According to the diagram, the first adult-like characteristic to appear in the tadpole are its hind legs. The forelegs, choice (4), appear later. Choices (1), (2) and (5) name structures that are only characteristic of the tadpole and are not found in the adult frog.

21. **Answer:** (2) The key phrase in the question is "must develop." Lungs are the only structures listed that must develop to make it possible for the adult frog to live on land. Choice (1), legs, are not necessary for living on land. Choices (3) and (4) do not develop, but are lost by the time the tadpole becomes an adult frog. Choice (5) is present in both tadpole and frog; it only changes shape.

22. **Answer:** (5) The question asks which kind of animal the tadpole is *most* like. Even though the tadpole is an amphibian, choice (3), it is most like a fish since it has gills, a caudal fin and a two-chambered heart. There is no real resemblance between the tadpole and choices (1), (2) and (4).

23. **Answer:** (3) According to the passage, the medulla oblongata controls the activities of the inner organs. Since the lungs are inner organs, it can be concluded that the medulla oblongata controls them.

24. **Answer:** (1) According to the diagram, hearing is controlled in the temporal lobe of the cerebrum. Choices (2) through (4) name other areas of the cerebrum. The cerebellum, choice (5), is another part of the brain.

25. **Answer:** (2) Since music is an artistic ability and the right side of brain is responsible for artistic ability, the right side of the brain must be responsible for a person's music ability. Choice (1) names the opposite side of the brain. Choices (3) through (5) do not name sides of the brain, but areas of the brain.

26. **Answer:** (2) According to the diagram, the occipital lobe of the brain is responsible for vision. The occipital lobe is located at the back of the head, so it could be concluded that a sharp blow to this region could affect a person's vision. None of the other choices names functions that the back part of the brain is responsible for.

27. **Answer:** (3) According to the passage, the cerebellum is responsible for balance. If a person loses control of his or her balance, then it might be concluded that there was something wrong with the cerebellum. None of the other choices names parts of the brain that control balance.

28. **Answer:** (1) According to the passage, the right half of the brain receives information from the left side of the body and vice versa. Therefore, information from the left eye is sent to the right side of the brain. Since the occipital lobe of the brain accepts information for vision, we can conclude that the right occipital lobe receives information from the left eye.

29. **Answer:** (3) In order for there to be a secondary immune response, there must have been a primary response leaving B cells ready to immediately produce antibodies the next time the disease attacks the body. Choices (1), (2) and (4) are irrelevant. Antibodies do not remain in the blood stream, choice (5), after a disease has been cured.

30. **Answer:** (2) Without B cells, antibodies could not be produced to fight off the virus. Choice (3) states the opposite effect. Choice (5) would become irrelevant if B cells could not multiply. Choices (1) and (4) make no sense in light of the question.

31. **Answer:** (5) A person contracts a disease because he or she has no antibodies against it. Choices (1) through (3) are in the body ready to start the immune response. The body has immune response defenses, choice (4), that have to be given time to take effect.

32. **Answer:** (1) Catching colds over and over again probably means there are many different viruses that cause colds. If all colds were caused by the same virus, choice (2), then it would be expected that the body would develop antibodies against the cold virus the first time it infects the body. Choice (3) is incorrect if choice (1) is correct. T cells, choice (4), do not make antibodies. Choice (5) is illogical.

33. **Answer:** (4) According to the chart, people with type O blood can donate their blood to anyone. In this context the phrase "universal donor" makes sense. Choice (2) is the opposite of the correct answer. Choices (1) and (5) are about receiving blood, not donating it. Choice (3) is true, but has nothing to do with donating blood.

34. **Answer:** (3) According to the chart, only people with type AB blood can receive blood of any type. People with the blood types mentioned in choices (1), (2) and (4) have few choices as to the kind of blood they can receive. Choice (5) is a genotype, not a blood type.

35. **Answer:** (5) According to the chart, 91.3% of North American Indians have type O blood. This is a much larger percentage than occurs in any other group listed on the chart.

36. **Answer:** (4) According to the chart, the percentage of people with AB blood is lower than for any other blood type. Choices (1), (2) and (5) name the other three blood types. Choice (3) names a genotype, not a blood type.

37. **Answer:** (3) The biome that occurs furthest north is the tundra. Since the tundra's climate is extremely cold and dry, this is what you would expect. Each of the other choices names another kind of biome.

38. **Answer:** (1) According to the map, a vast part of the midwestern states is made up of grasslands. Each of the other choices names a different kind of biome.

39. **Answer:** (5) Choices (1) through (4) can be eliminated because they name things that are not typical of a tropical rain forest biome. The climate of a tropical rain forest biome is described as wet most of the year. From this you could conclude that there is a large amount of rainfall in this biome.

40. **Answer:** (3) The key word in this question is "not." All choices except choice (3) name kinds of plants found in a tropical rain forest biome. Cacti are the only plants listed that are found in another type of biome, the desert biome.

Earth Science Review pp. 108–109

1. **Answer:** (2) This is correct because, according to the diagram, the temperature changes from 870° to 2,200° to 5,000°. The other choices are incorrect because they contradict the correct answer.

2. **Answer:** (4) This is correct because iron is listed as part of the composition of each layer. Choices (1) and (2) are incorrect because these elements appear only in the two top layers. Choice (3) is incorrect because aluminum appears only in the crust. Choice (5) is incorrect because nickel appears only in the outer and inner cores.

3. **Answer:** (5) This is correct because the crust is 32 km deep, compared to the rest of the earth, which is 6,468 km deep. The other choices are incorrect because they represent percentages that are much too large.

4. **Answer:** (4) This is correct because the crust consists of eight different elements. Choices (1) and (2) are incorrect because these layers consist of only nickel and iron. Choice (3) is incorrect because the mantle consists of only four elements. Choice (5) is incorrect because the layers are not similar in composition.

5. **Answer:** (5) This is correct because the temperature decreases to −55°C at the border between the troposphere and stratosphere; increases to 0°C at the border between the stratosphere and mesophere; decreases to −100°C at the top of the mesophere; then eventually climbs to 2,000°C at the top of the thermosphere. The other choices are incorrect because they do not account for the several increases and decreases that occur as altitude increases.

6. **Answer:** (3) This is correct because the temperature in the upper thermosphere ranges from 600°C to 2,000°C. Choice (1) is incorrect because this region is 1,000 km or higher above the earth. Choice (2) is incorrect because the ozone layer is located well below the thermosphere in the stratosphere and mesosphere. Choice (4) contradicts the fact that the temperatures in this region are very high. Choice (5) is actually true in some cases, but this cannot be inferred from the information given.

7. **Answer:** (2) This is correct because the diagram makes no reference to land that receives too much water. Choice (1) is incorrect because the diagram indicates that land that is snow-covered cannot be used for farming. Choices (3), (4) and (5) are incorrect because the diagram indicates that some land that is unsuitable for farming is too mountainous, under sea level or lacking in topsoil.

8. **Answer:** (4) This is correct because 30% of the land is already used for farming, and 20% is too dry for farming. Irrigating the dry land would result in making 50% of available land suitable for farming. The other choices are incorrect because they contradict the correct answer.

Chemistry Review pp. 126–127

1. **Answer:** (1) This is correct because the passage describes how molecular motion is different for gases, liquids and solids. Choices (2) and (5) are incorrect because no mention is made in the passage of temperature. Choices (3) and (4) are incorrect because no mention is made of molecular size.

2. **Answer:** (2) This is correct because, in order for ice to change into water and then into steam, the temperature must increase. Because the passage states that molecular motion is greatest for a gas and least for a solid, you can conclude that molecular motion must increase as temperature increases. Choice (1) is incorrect because it is opposite to the correct answer. Choices (3) and (4) are opposite to what can be inferred from the passage. Choice (5) is irrelevant.

3. **Answer:** (4) This is correct because molecules of a solid can only vibrate, while molecules of a gas are in constant, random motion and have no forces of attraction between them. No basis is given in the passage for choices (1), (2) and (3). Choice (5) contradicts the statement that gas molecules are in constant *random* motion.

4. **Answer:** (3) This is correct because gas molecules have more freedom of motion than liquid molecules. Choice (1) is incorrect because solid molecules, not gas molecules, exhibit vibratory motion. Choice (2) is not a good choice because, although gas molecules could conceivably collide with liquid molecules, this would not account for the gas bubbles rising to the top of the liquid. Choice (4) is incorrect because no mention is made of forces of attraction between liquids and gases. Choice (5) is incorrect because it is the reverse of the correct answer.

5. **Answer:** (3) This is correct because both the diagram and the explanation indicate that there is more space between ice molecules than between water molecules. Choices (1), (2) and (4) are incorrect because the diagram shows size, shape and number to be essentially the same for both ice and water. Choice (5) is irrelevant.

6. **Answer:** (2) This is correct because water is unusual in that it is more dense as a liquid (higher temperature) than as a solid (lower temperature). Since density decreases as substances expand, most substances must expand at higher temperatures. Choices (3) and (5) contradict the correct answer since they describe water, not most other substances. Choice (1) is incorrect because no mention is made of freezing points of substances. Choice (4) is incorrect because no mention is made of comparative densities except for water and ice.

7. **Answer:** (1) This is correct because water expands as it freezes. Choice (2) is incorrect because it is the spaces between molecules that increase in size, not the molecules themselves. Choice (3) is incorrect because the exact opposite is true. Choice (4) is incorrect because concrete has nothing to do with the information provided. Choice (5) is not even mentioned in the passage.

8. **Answer:** (5) This is correct because less dense substances will float on denser substances, and ice is less dense than water. Choice (2) is the opposite of the correct answer. Choices (3), (4) and (1) are not relevant to the information given.

Physics Review *pp. 151–152*

1. **Answer:** (2) This is correct because if a radioactive atom were to change size at all, it would get smaller as it releases subatomic particles and energy. Choices (1), (3) and (4) are incorrect because during the process of radioactive decay, energy and subatomic particles are released as a radioactive element changes into another element. Choice (5) is incorrect because all radioactive activity is accompanied by the release of gamma rays.

2. **Answer:** (2) This is correct because alpha particles are positively charged and beta particles are negatively charged. There is no evidence from the information given that any of the other choices would have an effect on these particles.

3. **Answer:** (5) This is correct because all types of radioactive decay are accompanied by gamma rays, and these rays are destructive to the cells of living things. The other choices limit the conditions under which exposure to radiation would be harmful, which contradicts the information provided.

4. **Answer:** (4) This is correct because in radioactive decay, the nucleus of an atom breaks down until a stable nucleus is reached. Choice (3) is incorrect because it is the opposite of the correct answer. Choice (1), (2) and (5) are irrelevant to the question.

5. **Answer:** (1) This is correct because the passage states that a fission reaction must be made to happen. Choices (2) and (4) are incorrect because a fission reaction is defined as the splitting of an atomic nucleus in which two smaller nuclei (matter) and energy are released. Choice (3) is incorrect because fission reactions are controlled in a nuclear reactor. Choice (5) is incorrect because the first fission reaction took place in 1938.

6. **Answer:** (3) This is correct because it is stated that large quantities of energy can be produced when many atoms undergo fission in rapid succession. Although choice (1) could be true, it is not mentioned in the passage. Choices (2) and (5) are incorrect because no mention is made of atomic size. Choice (4) is incorrect because there is nothing to suggest that the order of the reaction can be changed.

Instructions for Using This Answer Sheet

To be sure that your test results are properly scored and recorded:

- Use a soft lead pencil (not a pen) to mark your answers.
- Erase completely any errors or answers you wish to change.
- Make no unnecessary marks or calculations on this answer sheet or in your test booklet.
- Be sure the marks you make to fill in the circles are dark and fill the circle completely.

Do this: ●   Not this: ◉ ☒ ✓

DO NOT FOLD OR CREASE THE ANSWER SHEET.

In the time provided before you start the test, fill in the information in sections 1–10 on the answer sheet. In sections 9 and 10, be sure to write the letters or number in the box provided and mark the appropriate circle. (This helps avoid later scoring errors!)

Permission is granted to reproduce this form for student use.

Tests of General Educational Development

7 TEST BOOKLET NO. _____

8 TEST TAKEN AT _____

When completed this answer sheet must be treated as **Confidential Material.**

9 TEST FORM

☐

MN ○
MO ○
MP ○
MQ ○
MR ○
MS ○
MT ○
MU ○
MV ○
MW ○
MX ○
MY ○
MZ ○
SF ○
SG ○
SH ○
SJ ○
SK ○
SL ○
SM ○
AR ○
AS ○
MC ○
MH ○
LR ○
LS ○

10 TEST NUMBER

☐

① ② ③ ④ ⑤

TEST ANSWERS
Fill in the circle corresponding to your answer for each question. Erase cleanly.

DO NOT MARK IN YOUR TEST BOOKLET

1 ① ② ③ ④ ⑤	21 ① ② ③ ④ ⑤	41 ① ② ③ ④ ⑤	61 ① ② ③ ④ ⑤
2 ① ② ③ ④ ⑤	22 ① ② ③ ④ ⑤	42 ① ② ③ ④ ⑤	62 ① ② ③ ④ ⑤
3 ① ② ③ ④ ⑤	23 ① ② ③ ④ ⑤	43 ① ② ③ ④ ⑤	63 ① ② ③ ④ ⑤
4 ① ② ③ ④ ⑤	24 ① ② ③ ④ ⑤	44 ① ② ③ ④ ⑤	64 ① ② ③ ④ ⑤
5 ① ② ③ ④ ⑤	25 ① ② ③ ④ ⑤	45 ① ② ③ ④ ⑤	65 ① ② ③ ④ ⑤
6 ① ② ③ ④ ⑤	26 ① ② ③ ④ ⑤	46 ① ② ③ ④ ⑤	66 ① ② ③ ④ ⑤
7 ① ② ③ ④ ⑤	27 ① ② ③ ④ ⑤	47 ① ② ③ ④ ⑤	67 ① ② ③ ④ ⑤
8 ① ② ③ ④ ⑤	28 ① ② ③ ④ ⑤	48 ① ② ③ ④ ⑤	68 ① ② ③ ④ ⑤
9 ① ② ③ ④ ⑤	29 ① ② ③ ④ ⑤	49 ① ② ③ ④ ⑤	69 ① ② ③ ④ ⑤
10 ① ② ③ ④ ⑤	30 ① ② ③ ④ ⑤	50 ① ② ③ ④ ⑤	70 ① ② ③ ④ ⑤
11 ① ② ③ ④ ⑤	31 ① ② ③ ④ ⑤	51 ① ② ③ ④ ⑤	71 ① ② ③ ④ ⑤
12 ① ② ③ ④ ⑤	32 ① ② ③ ④ ⑤	52 ① ② ③ ④ ⑤	72 ① ② ③ ④ ⑤
13 ① ② ③ ④ ⑤	33 ① ② ③ ④ ⑤	53 ① ② ③ ④ ⑤	73 ① ② ③ ④ ⑤
14 ① ② ③ ④ ⑤	34 ① ② ③ ④ ⑤	54 ① ② ③ ④ ⑤	74 ① ② ③ ④ ⑤
15 ① ② ③ ④ ⑤	35 ① ② ③ ④ ⑤	55 ① ② ③ ④ ⑤	75 ① ② ③ ④ ⑤
16 ① ② ③ ④ ⑤	36 ① ② ③ ④ ⑤	56 ① ② ③ ④ ⑤	76 ① ② ③ ④ ⑤
17 ① ② ③ ④ ⑤	37 ① ② ③ ④ ⑤	57 ① ② ③ ④ ⑤	77 ① ② ③ ④ ⑤
18 ① ② ③ ④ ⑤	38 ① ② ③ ④ ⑤	58 ① ② ③ ④ ⑤	78 ① ② ③ ④ ⑤
19 ① ② ③ ④ ⑤	39 ① ② ③ ④ ⑤	59 ① ② ③ ④ ⑤	79 ① ② ③ ④ ⑤
20 ① ② ③ ④ ⑤	40 ① ② ③ ④ ⑤	60 ① ② ③ ④ ⑤	80 ① ② ③ ④ ⑤

Permission is granted to reproduce this form for student use.

POSTTEST Science

TIP Before you take this Posttest you may want to skim the test-taking Tips presented in our program. Use them to remind you of ways in which you can help yourself pass the GED test.

DIRECTIONS: Choose the one best answer for each item below.

Items 1–4 refer to the following passage and diagram.

If you touched something hot, you would instantly jerk your hand away. Such a response is called a reflex act. A reflex act involves a stimulus and a response. In this case heat was the stimulus. Moving your hand away was the response. During a reflex act, nerve impulses travel through a reflex arc that involves sensory neurons, the spinal cord and motor neurons. Heat was sensed by nerve endings in the skin, which caused nerve impulses to travel up sensory nerves to the spinal cord. Here associative neurons connected to motor neurons that caused the muscles of the hand and arm to contract, and you moved your hand away from the heat.

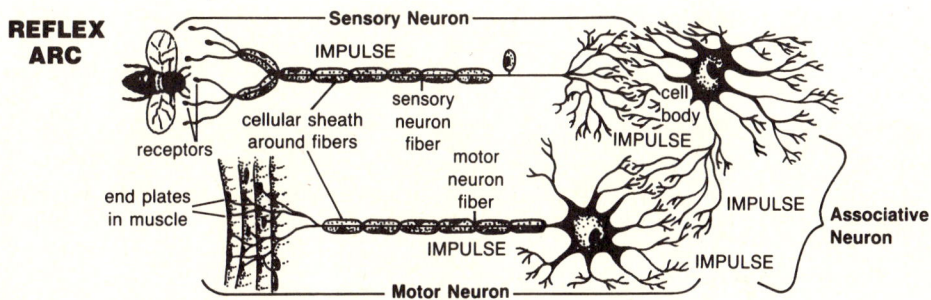

1. In a reflex arc, sensory impulses move

    (1) away from the stimulus
    (2) toward the stimulus
    (3) away from the spinal cord
    (4) away from the associative neurons
    (5) toward the muscle fibers

2. Which of the following would be a stimulus for a reflex act?

    (1) jerking your hand away
    (2) yelling "Ouch!"
    (3) being stung by an insect
    (4) blinking
    (5) swatting an insect

3. Which *best* describes the difference between sensory and motor neurons?

    (1) Both neurons carry nerve impulses.
    (2) Sensory neurons react to a stimulus; motor neurons cause a response.
    (3) Motor neurons react to a stimulus; sensory neurons cause a response.
    (4) Motor neurons carry impulses to the spinal cord; sensory neurons carry impulses away from the spinal cord.
    (5) Both are involved in the reflex arc.

4. In a reflex arc, associative neurons are responsible for

    (1) sensing a stimulus
    (2) causing muscles to contract
    (3) sending nerve impulses to sensory receptors
    (4) a motor response
    (5) transferring nerve impulses from sensory neurons to motor neurons

GO ON TO THE NEXT PAGE.

Items 5–8 refer to the following passage and diagram.

Every cell in the body is bathed and nourished by a fluid called lymph. Lymph is the liquid part of the blood that has passed, or diffused, out of the capillaries and into and around the cells. It consists of water, nutrients, white blood cells, various chemicals and cellular waste products. Lymph carries nutrients and oxygen to the cells and takes wastes and toxic substances away from them.

Some of the lymph is returned to the blood by passing directly back into the capillaries. Most lymph, however, diffuses into tiny lymph vessels that lie next to capillaries. As these lymph vessels become larger, they are called lymphatics. The lymphatics have thin walls and valves much like veins and are spread throughout the body. Lymphatics form still larger vessels called trunks. The trunks empty the lymph into two large ducts, the right lymphatic duct and the thoracic duct. Both ducts return the lymph to the blood by emptying into the large veins at the sides of the neck.

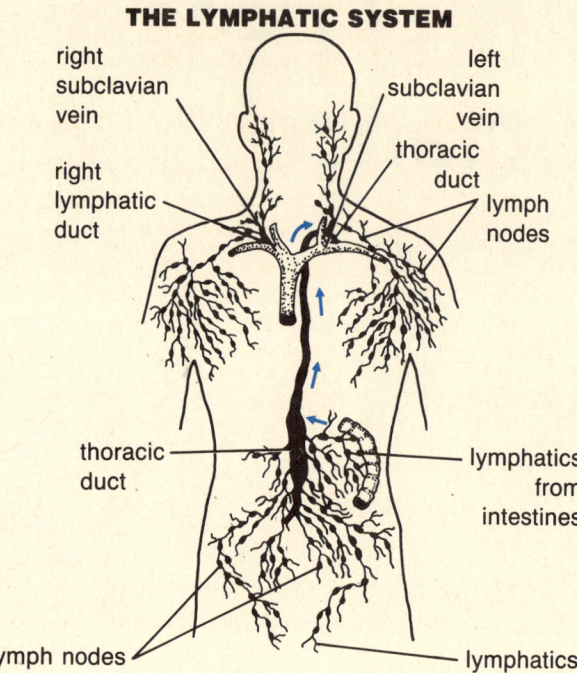

Enlargements at regular intervals along the lymphatics are called lymph nodes, or lymph glands. Active cells, among them white blood cells, in the lymph nodes filter out foreign material and kill any bacteria in the lymph. In this way the lymph nodes help purify the lymph before returning it to the blood.

5. According to the diagram, into what does the thoracic duct empty?

   (1) lymphatics from intestines
   (2) right lymphatic duct
   (3) right subclavian vein
   (4) left subclavian vein
   (5) lymphatics from the head and neck

6. Which of the following substances would be a source of harm to the body?

   (1) blood
   (2) nutrients
   (3) cellular waste
   (4) oxygen
   (5) white blood cells

7. According to the diagram, which lymphatic duct carries lymph from the right arm and the right side of the head and neck?

   (1) right lymphatic duct
   (2) right subclavian vein
   (3) thoracic duct
   (4) left subclavian vein
   (5) lymph nodes

8. Which of the following *best* describes the function of lymph nodes?

   (1) carry nutrients to cells
   (2) carry oxygen to cells
   (3) collect lymph
   (4) bathe tissues in lymph
   (5) purify lymph

GO ON TO THE NEXT PAGE.

Items 9–10 refer to the following paragraph and diagram.

Different classes of vertebrates have differently structured hearts. The heart of a fish, for example, is much simpler than that of a mammal. The diagram below compares the heart structure in five different classes of animals.

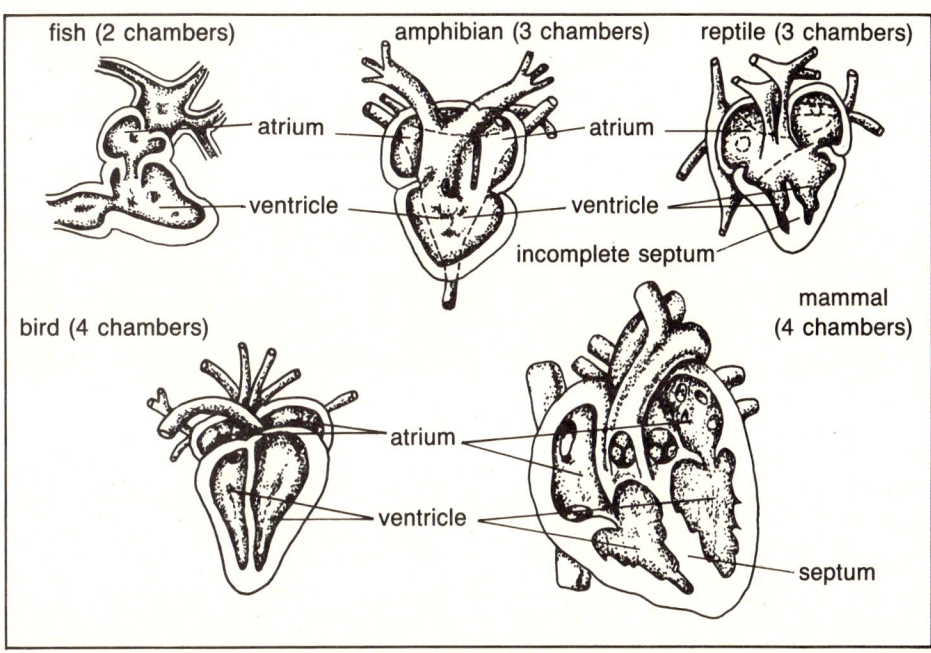

9. In which class of animals does the septum *first* appear?

   (1) fish
   (2) amphibians
   (3) reptiles
   (4) birds
   (5) mammals

10. When comparing the hearts in the diagram, the *most* striking difference is in the

    (1) shape
    (2) size
    (3) ventricles
    (4) atriums
    (5) number of chambers

Items 11–12 refer to the following passage.

If you watch a car drive past you, you know that the car is moving because it is changing position with respect to trees, telephone poles and other stationary objects. A system of landmarks and points against which motion is measured is called a frame of reference. A frame of reference must be chosen in order to describe how objects are moving in relationship to one another.

Whatever is taken as the frame of reference is usually assumed by the observer to be standing still. The most common frame of reference is the earth. When we say that a car is traveling at a speed 50 kilometers per hour, we are really saying that the car is traveling at a speed of 50 kilometers per hour relative to the earth.

It is interesting to consider an object's motion from more than one frame of reference. For example, suppose you are riding in a car going 70 kilometers per hour. A train passes you at a speed of 80 kilometers per hour. To an observer standing at the side of the road, the train appears to be moving quite fast. Yet to you the train appears to be moving rather slowly—because compared to the motion of the car in which you are riding, the train is traveling only 10 kilometers per hour.

GO ON TO THE NEXT PAGE.

11. A bus traveling at 25 mph is passed by a car traveling at 40 mph. Which of the following statements correctly describes the motion of the bus and the car?

   (1) Relative to the bus, the car is traveling 15 mph.
   (2) Relative to the car, the bus is traveling 65 mph.
   (3) Relative to the bus, the car is traveling 65 mph.
   (4) Relative to the car, the bus is traveling 25 mph.
   (5) Relative to the earth, both the bus and the car are traveling at the same speed.

12. When astronauts who traveled in space at a speed of 30,000 kilometers per hour were asked how it felt to move that fast, they said that they had no awareness of speed. Based on the information provided, which of the following statements could explain why they felt that way?

   (1) It is impossible to judge speed at high elevations.
   (2) The astronauts had no object or landmark with which to compare their motion.
   (3) The astronauts were too dizzy to accurately measure their speed.
   (4) Everything in outer space moves at speeds greater than 30,000 kilometers per hour.
   (5) Because they were moving in a circular orbit, the astronauts felt as if they were standing still.

Items 13–14 refer to the following diagram.

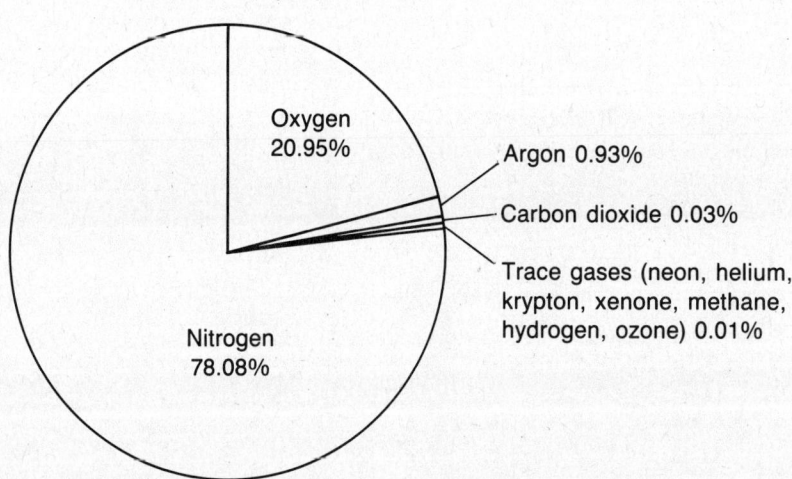

13. Which of the following information could *not* be determined from the diagram?

   (1) the amount of oxygen present in one liter of air
   (2) the amount of carbon dioxide present in the atmosphere compared to the amount of oxygen
   (3) the percentage of argon present in a one-liter sample of air if half of the argon has been removed
   (4) the amount of carbon dioxide present in a one-liter sample of air if one milliliter of carbon dioxide is added to the sample
   (5) the amount of hydrogen present in the atmosphere compared to the amount of nitrogen

14. A 200-ml sample of air is placed in a flask. A substance added to the flask reacts chemically with half of the nitrogen present to form a new substance. The amount of nitrogen remaining is

   (1) 39 ml
   (2) 78 ml
   (3) 100 ml
   (4) 156 ml
   (5) 200 ml

GO ON TO THE NEXT PAGE.

Items 15–18 refer to the following information.

Chemical reactions can be classified into general categories. Five of these categories are described below.

A. Synthesis: Two or more simple substances combine to form a new, more complex substance
B. Decomposition: A complex substance breaks down into two or more simpler substances
C. Combustion: A substance reacts with oxygen from the atmosphere to give off a large amount of energy in the form of heat and light
D. Single replacement: An uncombined element replaces an element that is part of a compound (A + BC → A + B)
E. Double replacement: Atoms in two different compounds replace each other; two compounds react to form two new compounds (AB + CD → AD + CB)

15. Propane, which is used as a fuel in portable stoves, burns in the presence of oxygen to produce large amounts of heat. This reaction is an example of

(1) synthesis
(2) decomposition
(3) combustion
(4) single replacement
(5) double replacement

16. Iron chloride ($FeCl_2$) reacts with potassium sulfide (K_2S) to form iron sulfide (FeS) and potassium chloride (KCl). This reaction is an example of

(1) synthesis
(2) decomposition
(3) combustion
(4) single replacement
(5) double replacement

17. When an electric current is passed through water (H_2O), oxygen (O_2) and hydrogen (H_2) are released. This reaction is an example of

(1) synthesis
(2) decomposition
(3) combustion
(4) single replacement
(5) double replacement

18. The active metal zinc (Zn) reacts with hydrochloric acid (HCl) to form zinc chloride ($ZnCl_2$) and hydrogen gas (H_2). This reaction is an example of

(1) synthesis
(2) decomposition
(3) combustion
(4) single replacement
(5) double replacement

Answers and Explanations

Science Posttest pp. 161–165

1. **Answer:** (1) According to the diagram and the passage, sensory impulses are set up by the stimulus and move away from it along sensory neurons to the spinal cord. Choice (2) states the opposite of the correct choice. Choices (3) and (4) would be toward the stimulus and not away from it. Motor impulses, not sensory impulses, move toward muscle fibers, choice (5).

2. **Answer:** (3) A stimulus causes a response or action. Only choice (3) would act as a stimulus. Each of the other choices would more likely be a response.

3. **Answer:** (2) The question asks what is the *difference* between the two types of neurons. Choices (1) and (5) both explain how sensory and motor neurons are alike. Choices (3) and (4) state incorrect comparisons.

4. **Answer:** (5) According to the diagram, associative neurons transfer the nerve impulses from sensory neurons to motor neurons. Choice (1) is accomplished by sensory nerves. Choice (2) is accomplished by motor nerves. Choice (3) is an incorrect statement. Choice (4) is accomplished by motor neurons.

5. **Answer:** (4) The diagram clearly shows that the thoracic duct empties into the left subclavian vein. Choices (1) and (5) do not name ducts. Choice (2) is the name of the other lymphatic duct. Choice (3) is the name of the vein into which the right thoracic duct empties.

6. **Answer:** (3) A foreign substance is any substance that must be removed from the lymph because it is toxic to the body. Cellular waste refers to substances that would be toxic if not removed from the body. Choices (2), (4) and (5) all name substances that are carried to the cells by the lymph. Choice (1) is not relevant since lymph is, in effect, part of the blood.

7. **Answer:** (1) The diagram clearly shows that the right lymphatic duct carries lymph from the right arm and the right side of the head and neck. Choices (2) and (4) name veins into which lymphatic ducts empty. Choice (3) names another lymphatic duct. Choice (5) names a part of the lymph system that is not a duct.

8. **Answer:** (5) The key phrase in the question is "*best* describes." The best description of the function of a lymph node that is listed is that it purifies lymph. Choices (1), (2) and (4) are not functions of lymph nodes. Though choice (3) happens in a lymph node, it is not the best description of a lymph node's function.

9. **Answer:** (3) The question asks when the septum *first* appears. The diagram clearly shows that the septum first appears in the reptilian heart as an incomplete septum. If the question had asked when a complete septum appears, then choice (4) would have been correct.

10. **Answer:** (5) The key phrase in this question is "the *most* striking difference." The diagram shows that the most striking difference is the number of chambers. All of the hearts are similar in shape, choice (1). Choice (2) would depend upon the size of the animal and size varies considerably within each class. All of the hearts have choices (3) and (4), even though the number may vary.

11. **Answer:** (1)   This is correct because the frame of reference is the bus, which is going 25 mph; from the bus the car appears to be going 15 mph (40 − 25 = 15). Choices (2) and (4) are incorrect because to the passengers in the car, the bus would actually appear to be traveling 15 mph backward, because the car is moving faster. Choice (3) is incorrect because the speeds are subtracted, not added. Choice (5) totally misses the meaning of the information given.

12. **Answer:** (2)   This is correct because the main idea of the passage is that it is necessary to use a frame of reference when describing the motion of objects. Choices (1), (4) and (5) are incorrect because nothing in the passage suggests that these factors affect the perception of velocity. Choice (3) could be true, but it does not pertain to the information provided.

13. **Answer:** (5)   This is correct because no specific percentage is given for hydrogen. Choices (1), (3) and (4) can be determined because the graph shows the percentage by volume of these gases. Choice (2) can be determined simply by forming a ratio of the percentages of the two gases.

14. **Answer:** (2)   This is correct because the amount of nitrogen present in two hundred milliliters of air is 156 ml, and half of this amount is 78 ml. The other choices are incorrect because they contradict the correct answer.

15. **Answer:** (3)   This is correct because the propane combines with oxygen to produce a significant amount of energy in the form of heat. Choice (1) is incorrect because two substances are not combining to form a new substance. (Heat is energy, not a substance.) Choice (2) is incorrect because no mention is made of propane being broken down into two simpler substances. Choices (4) and (5) are incorrect because there is no replacement taking place in this reaction.

16. **Answer:** (5)   This is correct because two compounds, iron chloride and potassium sulfide, are reacting to form two new compounds, iron sulfide and potassium chloride. Choice (1) is incorrect because two substances are not reacting to form one substance. Choice (2) is incorrect because a single substance is not breaking down. Choice (3) is incorrect because no combination with oxygen is indicated. Choice (4) is incorrect because this reaction involves two groups of atoms changing places, not one element replacing another.

17. **Answer:** (2)   This is correct because the water breaks down into its component elements, oxygen and hydrogen. The other choices are incorrect because water is breaking down, not combining with another substance.

18. **Answer:** (4)   This is correct because the element zinc replaces the element hydrogen in hydrochloric acid. Choice (1) is incorrect because two substances are not combining to form one new substance. Choice (2) is incorrect because a single substance is not breaking down into simpler substances. Choice (3) is incorrect because no mention is made of combination with oxygen or the liberation of energy. Choice (5) is incorrect because only one, not two, elements is being replaced.

Pretest/Posttest Diagnostic Chart

For the Pretest After you check your answers, look at the chart below. Circle the number of each problem you missed. Find the skills in which you need the most help. Then find those skills listed in the Table of Contents and begin your studying in those skill areas.

For the Posttest After you check your answers, look at the chart below. Circle the number of each problem you missed. Find the skills in which you still need help. Then find those skills listed in the Table of Contents and review those skill areas, paying special attention to the Strategies for Reading pages and the Answers and Explanations pages for the Practice items and the GED Mini-Test.

CONTENT AREAS

SKILLS	Life Science Pretest	Life Science Posttest	Earth Science Pretest	Earth Science Posttest	Chemistry Pretest	Chemistry Posttest	Physics Pretest	Physics Posttest
Identify Main Idea	1	1			19			
Restate Information	2, 3, 4, 7, 9, 10	3, 4, 5, 7	16					
Summarize Ideas	12	8			20			
Draw Conclusions	8, 5, 11	9, 10	15	13, 14			18, 22	11, 12
Use Given Ideas in Another Context						15, 16, 17, 18	17	
Identify Cause and Effect	6	2			14		21	
Recognize Unstated Assumptions				6	13			

168 PRETEST/POSTTEST DIAGNOSTIC CHART

Index

A-B-O system, 76
Acceleration, 132
Acid rain, 67, 97
Acids, 111, 122
Adaptive radiation, 32
Adenine (A), 60
Adrenal gland, 46
Air masses, 79, 89
Air pollution, 67, 97
Alimentary canal, 42
Alpha radiation, 151
Alveoli, 24
Ammonia, 25, 31
Amperes, 140
Amplitude, 144
Amylase, 42
Anal pore, 70
Anaphase, 16
Angle of incidence, 148, 149
Angle of refraction, 149
Angiosperms, 39
Animalia, 37, 38
Antarctica, 82
Antibiotics, 53
Antibodies, 49, 53, 75
Aorta, 44
Apical meristems, 35
Arteries, 44, 45
Arterioles, 45
Assumptions, 43
Atmosphere, 79, 87, 109
Atomic number, 116
Atoms, 111, 114
   bonding of, 115
   electrical charge, 137
   nuclear fission, 152
   radioactive, 151
Atria, 44
Australia, 32

Bacteria, 51
   as decomposers, 63
   as nitrogen fixers, 25
Bacterial disease, 51
   defenses against, 49
   immune response, 75
   substances that combat, 53
Bacilli, 51
Bases, 111, 122
B cells, 75
Beliefs, 50
Beta radiation, 151
Bile, 42

Biology. *See* Life science
Biomes, 77
Biosphere, 13
Blood type, 76
Boiling point, 112
Brain, 74
Brainstem, 74
Bronchi, 24
Bronchial tubes, 24
Bronchioles, 24
By-product, 21

Cambium, 35
Capillaries, 24, 45
Capsid, 52
Capsule, 51
Carbon, 31
Carbon dioxide
   in atmosphere, 87
   and photosynthesis, 21, 26
   and respiration, 23, 24, 26
   sources, 26
Carbon monoxide, 67
Carbon-oxygen cycle, 26. *See also* Photosynthesis; Respiration
Cause and effect, 29, 102, 138
Cells, 12, 14
   division of, 16, 60
   and immune response, 75
   origin of, 32
   of plants, 18
   reproductive, 17
   respiration in, 23
   See also Bacteria; Paramecium
Cellulose, 18
Cell wall, 18
Cenozoic Era, 72, 83
Cerebellum, 74
Cerebrum, 74
Chemical equations
   photosynthesis, 21
   respiration, 23
   Solvay process, 124
Chemical reactions, 119, 123
Chemistry, 111
   acids and bases, 122
   chemical reactions, 119, 123

   compounds, 115
   elements, 114
   matter, 112
   Periodic Table, 116
   solutions, 121
Chemotherapy, 53
Chlorophyll, 18, 21
Chloroplasts, 18
Chordata, 38
Chromatin material, 14
Chromosomes
   in bacteria, 51
   diploid and haploid number, 17
   formation of, 14, 16
   and DNA molecules, 60
Cilia, 70
Cinder cone volcanoes, 105
Circulatory system
   arteries and veins, 45
   heart, 44
   and internal respiration, 23
Cirque, 104
Classes
   of angiosperms, 39
   of vertebrates, 38
Coal, 98
Cocci, 51
Complementary pairs, 60
Composite volcanoes, 104
Compounds, 115, 123
Conclusions
   assess adequacy of data supporting, 120, 145
   distinguish from supporting statements, 131
Condensation, 91, 112
Coniferous forest, 77
Consequences, implied, 81
Consumers, 63
Context, 88, 113
Continental polar air mass, 89
Continental tropical air mass, 89
Contractile vacuole, 70
Cotyledons, 39
Covalent bonds, 115
Crest, of waves, 144
Crude oil, 98

Crust, of the earth, 80, 108
   faults in, 103
   rocks formed in, 84
Cystosine (C), 60
Cytoplasm, 14, 16

Dark reactions, 21
Darwin, Charles, 30
Data
   assess appropriateness of, 57
   assess adequacy of, 145
Daughter cells, 16
Deciduous forest, 77
Decomposers, 63
Democritus, 114
Denitrification, 25
Density, 127
Desert, 77
Dicots, 39
Dicotyledons, 39
Digestive system, 42, 49
Dinosaurs, 82, 83
Diploid number, 17
Direct respiration, 23
Diseases. *See* Infectious diseases
DNA, 60
   in bacteria, 51
   in viruses, 52
Dominant traits, 56, 58
Ducts, 42

Earth
   angle of sun's rays on, 90
   atmosphere, 87
   interior, 80
   surface, 80, 92, 101
Earthflow, 101
Earthquakes, 103, 106
Earth science, 78
   air pollution, 97
   atmosphere, 87
   earthquakes, 103
   erosion, 104
   fossil fuels, 98
   fossils, 82
   layers of the earth, 80, 108
   mass wasting, 101
   minerals, 85
   oceans, 92
   resources, 94, 96, 98

(Earth Science, cont.)
   resources, 94, 96, 98
   rock types, 84
   volcanoes, 105
   water cycle, 91
   weather, 89
Ecosystems, 13, 63, 77
Effect. See Cause and effect
Electric charge, 137
Electric circuit, 129, 140
Electricity
   charge, 137
   electric current, 139, 140
   and magnetism, 141, 142
   wind-generated, 96
Electric current, 129, 139
   and electric circuits, 140
   and magnetism, 141
Electric field, 137
Electromagnetism, 141, 142
Electromagnetic waves, 147, 151
Electromagnets, 141
Electrolysis, 119
Electron cloud, 114
Electrons, 114
   bonding, 115
   charge, 137
   movement, 139, 140
Elements, 111, 114
   atomic number, 116
   in layers of the earth, 80, 108
   in organic molecules, 31
   radioactive, 129, 151
Embryo, 28
Embryonic similarity, 28
Endocrine glands and system, 46
Endoplasmic reticulum (ER), 14
Endothermic reaction, 123
Energy
   in chemical reactions, 119
   conservation of, 130
   from earthquakes, 103
   and ecosystems, 63
   from nuclear fission, 152
   and radioactivity, 151
   resources, 94, 96, 98
   in waves, 144
   and work, 134
Entropy, 111, 123
Environment, 12
Enzymes
   in digestive system, 42
   in photosynthesis, 21
Epicenter, 103
Epidermis, 49
Era, 78

Erosion, 79, 104
Esophagus, 42
Evaporation, 71, 91
Evolution, 28
   and genetic drift, 33
   and geologic time, 72
   and natural selection, 30
   and speciation, 32
   and taxonomy, 37
Evolutionary theory, 28, 30
Exothermic reaction, 123
Explicit main idea, 15
External respiration, 23
Eyes, pollutants that damage, 67

Fact, 95
Farming, land suitable for, 109
Fault, geologic, 103
Fermi, Enrico, 152
F_1 and F_2 generations, 56
Fish, respiration in, 23
Flagella, 51
Focus, of earthquakes, 103
Food chain, 63, 79, 99
Food vacuole, 70
Force, 129
   electrical, 137
   magnetic, 142
   and motion, 132
   resultant, 133
Fossil, 82
Fossil fuels, 98
   pollution due to, 67, 97
   vs. wind energy, 96
Freezing point, 112
Frequency, 144
   of light waves, 147
Frogs, 73, 99
Frontal lobe, 74
Fungi, 37, 63

Gamma rays, 151
Gametes, 17
Gases, 112
   air pollutants, 67, 97
   in atmosphere, 87
   molecular motion, 126
   in solution, 121
   sound waves in, 146
Gasoline engine, 134
GED Mini-Tests
   acid rain and smog, 67–68
   chemical reactions, 123–24
   defenses against disease, 53–54
   DNA molecule, 60–61
   earth science, 84–85
   electromagnetism, 141–42
   endocrine system, 46–47

   evolutionary theory, 32–33
   fossil fuels, 98–99
   heat and work, 134–35
   light waves, 148–49
   nitrogen and carbon-oxygen cycles, 25–26
   Periodic Table, 116–17
   plant cells and trees, 18–19
   seed plants, 39–40
   volcanoes, 105–6
   water cycle, 91–92
Generalizations, 81
Genes
   and blood type, 76
   DNA molecules, 60
   homo- and heterozygosity, 59
   represented in Punnett square, 58
Genetic drift, 33
Genetic crosses, 56, 58
Genetics, 13, 56
   DNA molecules, 60
   genotype and phenotype, 58
   homo- and heterozygosity, 59
Genotype, 58
   and blood type, 76
Geologic time, 72, 83
Gills, 23
Glaciers, 104
Glands, 42, 46
Glucose
   and photosynthesis, 18, 21, 26
   and respiration, 23, 26
Golgi apparatus, 14
Gonads, 46
Grasslands, 77
Green plants, 63
Ground water, 71
Guanine (G), 60
Gullet, 70
Gymnosperms, 39

Haploid number, 17
Heart, 44
Heat energy, 119, 130
Hemispheres, of brain, 74
Hereditary, defined, 12
Hertz, 144
Heterozygous, 59
Homologous structure, 28
Homozygous, 59
Hormones, 46
Horn, 104
Host cell, 52
Human digestive system, 42
Human population density, 66
Hybrids, 56, 59
Hydrochloric acid, 42
Hydrogen, 21, 31, 122

Ice, 127
Igneous rock, 84
Immune response, 75
Implication, 36
Implicit main idea, 15
Incidence, angle of, 148, 149
Inclined plane, 135
Indirect respiration, 23
Inertia, 132
Infectious diseases, 13, 49
   bacterial, 51
   defenses against, 49
   immune response, 75
   substances that combat, 53
   viral, 52
Inference, 81
Inner core, of earth, 80, 108
Internal respiration, 23
Internodes, 35
Interphase, 16
Intestines, 42
Invisible waves, 147
Ionic bonds, 115
Ions, 32, 122
Iron, 80

Kangaroo, 32
Kinetic molecular theory, 126
Kingdoms, 37
Koala bear, 32

Land, suitable for farming, 109
Landslide, 101
Large intestine, 42
Lava, 84, 105
Law, scientific, 130
Law of conservation of energy, 130
Law of reflection, 148
Leaves, 39
Legumes, 25
Leucoplast, 18
Life, origin of, 31
Life science, 12
   brain, 74
   cells, 14, 16, 17, 18, 70
   circulatory system, 44, 45
   ecosystems, 63, 77
   endocrine system, 46
   evolution, 28, 30, 31, 32, 34, 72
   genetics, 56, 58, 59, 60, 76
   human digestive system, 42
   infectious diseases, 49, 51, 52, 53, 75
   metamorphosis, 73
   nitrogen cycle, 25
   photosynthesis, 21, 26
   plants, 35, 39
   populations, 65, 66

respiration, 23, 24, 26
taxonomy, 37, 38
water cycle, 71
Light reactions, 21
Light waves, 147, 148–49
Liquids, 112
   molecular motion, 126
   sound waves in, 146
Liver, 42
Logical fallacies, 64
Longitudinal waves, 146
Lungs, 23, 24
   and circulatory system, 44
   pollutants that damage, 67
   mucus lining, 49
L waves, 103
Lymphocytes, 49, 75
Lysosomes, 14

Machines, 135
Macronucleus, 70
Macrophage, 49
Magma, 84, 105
Magnetic field, 142
   of earth, 80
   electric, 141
Magnetism, 141, 142
Magnetite, 142
Main idea, 15
Mantle, 80, 84, 108
Maritime polar air mass, 89
Maritime tropical air mass, 89
Marsupials, 32
Mass wasting, 79, 101
Matter, 111, 112
   density, 127
   kinetic properties, 126
   and radioactivity, 151
Mechanical energy, 130
Medium, for sound waves, 146
Medulla oblongata, 74
Meiosis, 17
Melting point, 112
Mendel, Gregor, 13, 56
Mendelian genetics, 56
Metamorphic rock, 84
Metamorphosis, 73
Metaphase, 16
Methane, 31
Mesosphere, 87, 109
Mesozoic Era, 72, 83
Metals, 116
Micronucleus, 70
Middle lamella, 18
Mineral crystals, 84
Minerals, 85
Mitochondria, 14
Mitosis, 16
Molecular motion, 130
Molecules, 111, 115
   kinetic molecular theory, 126
   and sound waves, 146
Monocots, 39

Monocotyledons, 39
Motion
   laws of, 130, 132
   and resultant force, 133
Mouth, 42
Mucus, 42, 49
Mucus membranes, 49
Mudflow, 101

Natural gas, 98
Natural selection, 30
Neutrons, 114, 137
Newton's laws of motion, 130, 132
Nitric oxide, 67
Nitrification, 25
Nitrite, 25
Nitrogen, 25, 31, 60, 87
Nitrogen dioxide, 67
Nitrogen fixation, 25
Nodes, 35
Nomads, 65
Nonmetals, 116
Nonrenewable resources, 79, 94
Nuclear chain reaction, 152
Nuclear fission, 152
Nuclear membrane, 14, 16
Nuclear physics, 151, 152
Nuclear reactor, 152
Nucleus
   of atom, 114, 151
   of bacteria, 51
   of cells, 14
Nucleolus, 14
Nucleoplasm, 14
Nucleotides, 60
Nutrients, digestion of, 42. See also Food chain

Occipital lobe, 74
Oceans, 92
Oersted, Hans Christian, 141
Offspring, 13
Ohms, 140
Oil. See Petroleum
Oparin's theory, 31
Opinion, 95
Ores, 85
Organelles, 14, 18
Organic molecules, 31
Organs
   in digestive system, 42
   of respiration, 23
Organ systems, 13
Outer core, 80, 108
Ovaries, 46
Oxygen
   in atmosphere, 87
   and organic molecules, 31
   and photosynthesis, 21, 26

   and respiration, 24, 26
Ozone, 87, 109

Paleozoic Era, 72, 83
Pancreas, 42
Paramecium, 70
Parathyroid gland, 46
Parent cell, 16, 17
Parietal lobe, 74
Pathogens, 49
Pectin, 18
Pellicle, 70
Pepsin, 42
Periodic Law, 116
Periodic Table, 111, 116
Petroleum, 79, 94, 98
Phagocytes, 49
Phase change, 112
Phenotype, 58
Photosynthesis, 13, 21
   vs. respiration, 23, 26
   role of chlorophyll in, 18
pH scale, 122
Physics, 128
   electricity, 137, 139, 140, 141
   magnetism, 141, 142
   motion and force, 132, 133
   nuclear, 151, 152
   thermodynamics, 130
   waves, 144, 146, 147, 148–49
   work, 134, 135
Pituitary gland, 46
Plant cells, 18, 21
Plants, 35, 39
Plasmids, 51
Plastids, 18
Pollution, 67, 97
Pons, 74
Population
   and blood type, 76
   density, 66
   growth, 65
Precipitation, 71, 91
Primary consumers, 63
Primary immune response, 75
Primary walls, 18
Primary waves (P waves), 103
Producers, 63
Prophase, 16
Protein, 14, 25
Protons, 114, 116, 137
Pulley, 135
Punnett square, 58
Purebred strains, 56
P waves, 103

Radioactive decay, 151, 152
Radioactivity, 129, 151
Radiation, 151
Rain. See Acid rain; Precipitation
Rays, 148

Recessive traits, 56, 58
Reflection, 129, 148
Refraction, 129, 148–49
Resistance, electrical, 140
Resources, 94, 96, 98
Respiration, 13, 23, 24, 26
Resultant force, 133
Ribosomes, 14
Richter scale, 103
Rocks, 82, 84
Root caps, 35
Roots, 35

Salinity, 92
Saliva, 42
Salts, 111
   from acids and bases, 122
   elements that form, 116
   in the ocean, 92
Scientific law, 130
Secondary consumers, 63
Secondary immune response, 75
Secondary walls, 18
Secondary waves (S waves), 103
Sedimentary rock, 84
Seed plants, 39
Septum, 44
Seismic waves, 80, 103
Shape of waves, 144
Shield volcanoes, 105
Single-celled organisms
   bacteria, 51
   kingdom of, 37
   respiration in, 23
Skin, 49
Small intestine, 42
Smelting, 85
Smog, 67
Sodium chloride, 92, 122
Sodium bicarbonate, 123, 124
Soil creep, 101
Solar energy, 94
Solids, 112
   molecular motion, 126
   sound waves in, 146
Solubility, 121
Solute, 121
Solution, 111, 121
Solvay process, 124
Solvent, 121
Sound waves, 146
Speciation, 32
Species, 13
   of seed plants, 39
   taxonomy, 37
Spindle, 16
Spirilla, 51
Starch, 18
Stomach, 42
Strategies for reading
   assess adequacy of data to support conclusions, 120, 145

INDEX 171

(Strategies for reading, *cont.*)
  assess appropriateness of data, 57
  distinguish conclusions from supporting statements, 131
  distinguish fact from opinion, 95
  identify an implication, 36, 81
  identify cause and effect relationships, 29, 102, 138
  identify logical fallacies in arguments, 64
  identify the main idea, 15
  recognize unstated assumptions, 43
  recognize the role of values in beliefs, 50
  restate information, 22
  transfer concepts to new context, 88
  use given ideas in another context, 113
Stratosphere, 87, 109
Sugar, 60, 115. *See also* Glucose
Sulfa drugs, 53
Sulfur, 97
Sulfur dioxide, 67
Sulfur oxides, 97
Sulfuric acid, 67, 122
Sun
  angle of rays on earth, 90
  as energy source, 63, 94
  role in photosynthesis, 21
  and water cycle, 71
Supporting statement, 131
Surface waves (L waves), 103
S waves, 103

Tadpole, 73
Talus slope, 101
Tartrate, 123
Taxonomy, 37
T cells, 75
Telophase, 16
Temperature
  of layers of atmosphere, 87, 109
  of layers of the earth, 80, 108
  of oceans, 92
  and solubility, 121
  and speed of sound, 146
Temporal lobe, 74
Tertiary consumers, 63
Testes, 46
Thermodynamics, 128, 134
  laws of, 130
Thermosphere, 87, 109
Thymine (T), 60
Thymus gland, 46
Thyroid gland, 46
Trace fossils, 82
Trachea, 24
Traits, 56, 59
Transpiration, 71
Transverse waves, 147
Trees, 19, 39
Trichocysts, 70
Tropical rain forests, 77
Troposphere, 79, 87, 109
Trough, of waves, 144
Tundra, 77

Uranium, 151

Vaccines, 53
Vacuoles, 14, 70
Values, 50
Vaporization, 112
Vectors, 129, 133
Veins, 44, 45
Velocity, 144
Vena cava, 44
Vent, 105
Ventricles, 44
Venules, 45
Vertebrates, 38
Virions, 52
Viral diseases, 52
  immune response, 75
  substances that combat, 53
Viruses, 49, 52
Volcanoes, 84, 105, 106
Voltage, 139, 140
Volts, 139

Water
  as a compound, 115
  density changes, 127
  pH, 122
  release in respiration, 23
  role in photosynthesis, 21
Water cycle, 71, 79, 91
Water table, 71
Water vapor, 31, 71, 87, 91
Waves, 129, 144
  electromagnetic, 147
  seismic, 103
  of sound, 146
Wavelength, 144
Weather, 89
Wind, 96
Windmills, 96
Work, 134

Zygote, 17